D0463414

NEW YORK INSTITUTE
OF TECHNOLOGY LIBRARY

Originality and Competition in Science

Originality and Competition in Science

A Study of the British High Energy Physics Community

Jerry Gaston

Foreword by John Ziman

The University of Chicago Press

Chicago and London

JERRY GASTON is associate professor of sociology at Southern Illinois University. He is a specialist in the sociology of science and has published a number of articles in scholarly journals.
[*1973*]

The University of Chicago Press, Chicago 60637
The University of Chicago Press, Ltd., London

Published 1973
Printed in the United States of America
International Standard Book Number: 0-226-28429-8
Library of Congress Catalog Card Number: 73-81313

To Mary Frank

Contents

Foreword by John Ziman ix

Acknowledgments xvii

1. Introduction 3
2. The Organization of Basic Research 12
3. Research in High Energy Physics 21
4. The Reward System 32
5. Competition in Science 69
6. Competition in High Energy Physics 94
7. Communication in the High Energy Physics Community 130
8. Summary and Conclusions 159

Appendix:
Prestige, Productivity, and Recognition Indexes 175

Notes 183

Bibliography 195

Index 207

Foreword

This book about my neighbors, the *uk-hep* and *uk-het* clans, contains much interesting information that will be used by other anthropologists in their analysis of our peculiar culture. Yet there are many aspects of clan life and work that would not be immediately apparent to a foreigner who had not been brought up in our native ways. This is particularly difficult in the case of the hep and het clans, whose rituals are performed to the accompaniment of songs in a cryptic symbolic language whose meaning cannot be translated to the uninitiated. As an *uk-solstath,* however, I have had the good fortune to learn some of these songs for use in the rituals of my own clan, and have also had many opportunities to observe the uk-heps and uk-hets in their daily lives. I therefore felt it my duty to the science of anthropology, in commending this book to you, to sketch in some of this background as seen by one of the natives.

The first point to make clear is that all members of these clans have dedicated their lives, by a series of magical ceremonies of great psychological weight, to a single purpose—the cultivation of the *elpar* fruit (*nucleonicus fermii*). Whatever secular power or worldly wealth may accrue to them from public success in this subtle and highly competitive art, is almost irrelevant beside the fascination of horticulture itself. Those who have been called away into the very highest ranks of chieftainship in the tribe as a whole, whether in warlike or peaceful pursuits, always speak with genuine regret of the laborious days and nights they once spent tending the young plants, controlling the irrigation channels, examining the harvest of husks, and selecting the sweetest smelling kernels. Despite the unworthy spirit of competition that occasionally seizes upon individuals and groups when some new variety is coming to seed or a new crop is ready for picking, the normal attitude is one of friendliness and cooperation between all members of the clans. The social psychology of clan life puts great emphasis on such fraternity, to the detriment of good relations between different members of different clans. All heps and hets of whatever tribe firmly believe that the cultivation of the elpar (which is, incidentally, quite inedible) is the most important thing in life, and that all the tribal resources should be devoted to this single end. Personal rivalries within a clan are quite insignificant by comparison with this fiercely sectarian attitude. This clannishness even extends across tribal boundaries. As we shall see in a moment, the hep and het clans are so distinct in their social roles that one may find closer fraternity between, say, uk-hets, us-hets, eur-hets, and russ-hets, even though these belong to different tribes that are often at war with one another, than between the uk-hets and uk-heps alongside whom they actually live.

The differentiation of role between heps and hets arises quite naturally out of the peculiar life cycle of the elpar. The reproductive phase, in which the seeds germinate rapidly, grow to young plants, flower, and produce a few new seeds, is under the charge of the hets; the fruiting phase, which lasts much longer, and leads to a large crop, is the care of the heps. The horticultural activities appropriate to these two phases are so entirely different that they lead to entirely different styles of social organization within the two clans. Between solstaths and solstaphs we also have a similar differentiation, but the distinction between the two clans is not nearly so marked as it is between hets and heps, which stand at almost opposite poles in the horticultural temperament.

The sole aim of the het is to breed a significantly new variety of elpar, as measured by the beauty of its flower and eventually by the fragrance of its fruit. This is essentially a solitary task, demanding no more than a few seed trays and pots, a little fertile compost, and a modest supply of water, light, and air. The secret of success is thought to lie mainly in the choice of ritual chants, which must be sung with perfect accuracy and clarity as the seed germinates and as the young plant pushes out its leaves and flower buds. The hets are, incidentally, the supreme masters of the symbolic language, and often vie with one another in the composition of more and more sophisticated and esoteric songs. But the individual het usually works alone, or in voluntary collaboration with two or three of his fellows, thus avoiding the dangers of mistakes arising from mutual misunderstandings as to the text actually being chanted. Such mistakes are especially prevalent when the attempt is made to breed new plants on a large scale, using numerous junior assistants to sing standardized songs that they do not fully understand. Hets cooperate mainly in the exchange of seeds, and of the appropriate songs, or in occasional sessions of mutual fertilization of their flowering plants, for the elpar quickly becomes self-sterile after a few generations of inbreeding. Such exchanges are indeed so essential that many young hets, lacking confidence in their own stocks of seed, form cliques of "clients" around their more notable and experienced colleagues, raising minor variants of any new variety that may thus be handed them. But such informal social groupings are essentially voluntary, and a het patron has no enforceable authority over his clients.

The heps, on the other hand, are tightly organized in

distinct groups. The members of each group work together as a gang, under the direction of an experienced headman. The harvesting of a satisfactory crop of elpar is extremely expensive and laborious, calling for great communal resources and efficient planning of specialized tasks. The enormous irrigation works that supply water continuously over a large area of fertile land are now no longer the sole responsibility of the hep clan that originally planned and built them, and are shared by many such gangs, but much labor is still required to prepare the land for each new crop. The young plants—usually acquired from friendly hets—must be set out according to a strict ritual pattern, and watered, weeded, pruned, etc., over a period of several years until the fruit has formed. But only a small proportion of the vast crop of husks actually contains a ripe kernel. One of the most exacting duties of a member of a hep gang is to oversee the work of several scores of slave women who break open each husk and look for the tiny kernels. These, in turn, must be closely scrutinized for signs of decay or other blemishes before they can be added to the accumulated harvest. At every stage, therefore, the work is arduous, time-consuming, and highly specialized. Each member of the gang must know his own job and carry it out with perfect precision in coordination with his fellows, if the whole crop is not to be a failure. The responsibilities hang heavily on the headman, especially since he must, in his turn, subordinate the work of his gang to the overall irrigation program of the tribe.

The social organization of the hep clan is thus naturally hierarchical, with authority in the hands of the caste of headmen. Rivalry between individuals within a gang is firmly suppressed, and competition between gangs in the same tribal region must be controlled. The competitive incidents reported by anthropologists have almost always involved hep gangs from different tribes—uk-heps in competition with us-heps, for example. Yet even these tensions are repressed at the great corroborees of het and heps, where seeds and plants, ritual chants and fruit, are compared, judged, and exchanged in a great orgy of friendship and mutual congratulation. Anthropological studies that concentrate on the occasional breakdown of these norms miss the most significant features of our clan system.

The most subtle aspect of this system is the symbiotic relationship between the hets and the heps. At first sight one would suppose that the heps are clearly subordinate, their task being simply to bring into production each newly bred variety. But the truth is more complex. It is often argued, for example, that it is the hets who have the more lowly role; theirs is the effeminate task of growing a few seed plants for the manly, physically active hep gangs who do the real work. The most thoughtful hets themselves emphasize that the quality of a new variety cannot be judged solely by the beauty of the flowers, but must eventually be tested by the fragrance and quantity of the fruit. In some cases, such testing takes many years —too long a time for those impatient hets who practice a program of random crossing without selection, hoping that a valuable new variety may occur by chance among the many seeds they produce.

But these subtleties are well understood by the leading members of the clans, and the award of high status amongst uk-hets by no means goes to those claiming to have produced the most new varieties of seed. In particular the rank of Fel-roy—a title of nobility that completely outweighs all other measures of social standing among the clans of uk—is given to an uk-het only for the breeding of some outstanding new variety that has been thoroughly tested in cultivation. The fact that this rank (for which there is no real equivalent amongst the us-hets and us-heps) is occasionally won at an early age shows how much more weight is given to quality than to quantity of production.

Amongst the uk-heps, although some of the leading headmen are Fel-roys; effective rank is much more closely linked to authority over a gang. A headman with a gang of sturdy heps working in a high-flow irrigation system can acquire great standing in the clan, not so much by his own personal achievements as by the labors of his gang. But the assignment of public rank amongst the members of such a closely organized group presents may obstacles, due to the specialized division of labor and the long time needed to harvest each crop. In practice, therefore, the judgment of the headman concerning the ability of each member of his gang is paramount. This is one of the key features of the hep clan as a social institution—a characteristic that contrasts strongly with the artistic individualism favored by the hets.

Our clans are not, of course, hereditary in the strict sense but are recruited by adoption. In the land of uk it is universally believed that the trait of *wrangler*—loosely translated "Green Fingers"—although rare among the people as a whole, may infallibly be detected at an early age by magical incantations in the ritual language. Young men and women in whom this highly prized quality has been observed are then given special horticultural training by which their natural ability is supposedly strengthened. A further ceremonial test in their twenty-second year decides their fate. A few of the most superior wranglers are then adopted by one or other of the most experienced hets, since green fingers are obviously much more valuable in the raising of delicate new varieties than in the large-scale production of a crop. The next grades of initiates are adopted into the hep clan—and retain throughout their lives a lingering jealousy of the hets whom they once aspired to emulate. But the training of heps is so specialized and toilsome that they are seldom able to learn the more powerful ritual songs and are soon cut off from the skills of seed raising. This explains both the division of labor between the clans and, to some extent, the subtle feelings of superiority and inferiority between them.

But to understand the system in all its dimensions one must go beyond the static picture presented by many contemporary anthropologists, who seem to suppose that the same social institutions have existed from time immemorial. The uk-het and uk-hep clans date back to the beginning of this century, to the great headman JoJo and his adopted son Rŭth, who was himself the headman or adopted father of many of the older headmen of our day. Rŭth himself was what we should now call a hep, but he had an unusual knack for raising many sturdy new varieties of plant without the aid of ritual songs, and was indeed somewhat scornful of many of the more fanciful varieties being bred by the hets of his day. This attitude also, transmitted via his adopted sons, is by no means outmoded amongst the present generation of uk-heps, who still maintain among themselves something of the family feeling of the founder of their clan.

I must apologize to readers for this nonscientific description of the psychological and historical aspects of our tribal life, but it seemed that many of the observations recorded in these anthropological researches would lose their significance unless seen in this context. As a humble member of a neighboring clan who has had the good fortune to become acquainted with several visiting anthropologists, I felt that it might be of some use to say some of these things in my own simple fashion.

University of Bristol *John Ziman*

Acknowledgments

This book is based on my Ph.D. thesis at Yale University. The research for this project could not have been completed without the assistance provided by a large number of people, and I want to express my appreciation to them. For assistance in deciding to undertake this particular project and the financial assistance to carry it out, I am indebted to Professor Diana Crane and Professor Wendell Bell. Professor Crane sparked my interest in the sociology of science, and the possibility of doing my thesis research in Britain was encouraged by Professor Bell. Both are responsible for obtaining the necessary funds that provided travel and other expenses: Professor Crane obtained an NSF Thesis Improvement Grant for me and continued to advise me from her position at Johns Hopkins University, and Professor Bell obtained support from the Yale Committee on Comparative and European Studies and from the Sociology Department Graduate Student Fund.

For additional encouragement at the beginning of the research, I am grateful to Professor A. B. Hollingshead and Professor Burton Clark. From the Department of the History of Science and Medicine, I was fortunate to have the wise counsel of Professor Derek J. de Solla Price for whose continued interest I am grateful.

At the request of Professor Bell, Professor Julius Gould, Head of the Department of Sociology at the University of Nottingham, graciously allowed me to be affiliated with his department as a research associate. Professor Gould advised me on several important points, arranged for me to meet certain people, and offered the facilities of his department and university to me. I also want to acknowledge the advice and suggestions of Mr. Michael King of the University of Nottingham Sociology Department.

I was somewhat uneasy about visitations to the national laboratories, but through the hospitality of Mr. William Woodall, Chief Administrative Officer of the Rutherford High Energy Laboratory, and Mr. Jack Wyatt, of the Daresbury Nuclear Physics Laboratory, the visits to these laboratories were some of the more pleasurable aspects of my stay in Britain. My wife and I stayed for seven weeks at Cosener's House (RHEL) and Norton Lodge (DNPL). These lovely surroundings did much to make my task enjoyable.

At the Rutherford lab, I am especially grateful to Dr. G. H. Stafford, Deputy Director, for his hospitality. Mr. David Salter and Mr. Peter Nichols also provided me with information about any questions I had.

At the Daresbury Nuclear Physics Laboratory, Professor A. W. Merrison (the Director of the laboratory and now Vice-Chancellor of Bristol University) was very generous with his time and did everything possible to make my visit successful.

Upon my return to the Yale Department of Sociology, Professor Anthony Oberschall agreed to assist me in the final stages of this project. It is a difficult task to step into a project without having been involved from the start, and I am especially grateful to Professor Oberschall for his time and the many suggestions and contributions that he made.

I am especially grateful to the scientists in the United

States who gave their time for preliminary interviews and to the British high energy physicists who spent from slightly less than an hour to over three hours with me in formal interviews and sometimes much longer over tea, coffee, or lunch.

I want to thank my colleagues, Herman Lantz and Charles Snyder, for their moral support and for providing the departmental resources necessary to complete the revisions and final manuscript preparations. I thank especially Tia Herring, Pat Landis, and Ruth Malek for their assistance in typing the manuscript.

I also thank Professor Edward Shils, whose pertinent criticism on earlier drafts enhanced the quality of the book. None of the persons who assisted me has any responsibility for the errors I made.

The most appreciation goes to my wife, Mary Frank, who worked at every stage of this research. She transcribed the preliminary interviews, wrote the letters requesting interviews, transcribed the one hundred and eighty recorded interviews, coded all the interviews once, and typed the preliminary and final copies of the thesis. Without her help and encouragement during difficult times, this research would not have been the pleasure it was, and would have required at least twice the time.

Some of the materials in this book have appeared in my articles (and where necessary permission to use them here has been given) as follows: "The Reward System in British Science," *American Sociological Review* 35 (1970): 718–32; "Secretiveness and Competition for Priority of Discovery in Physics," *Minerva* 9 (1971): 472–92; and "Communication and the Reward System of Science: A Study of a National 'Invisible College'," pp. 25–41 in *Sociological Review Monograph,* edited by Paul Halmos, 1972.

Originality and Competition in Science

Introduction 1

Science develops to a large extent on the accumulation of discoveries that validate or modify previous thinking about nature. Scientists must contribute to the store of knowledge or they are not valuable to the scientific community.[1] If a scientist cannot contribute something original or work with someone on something original, he might just as well choose another profession.

The theme of my book is originality and competition in science. Original research consists of not just doing something on a topic that no one has ever worked on before, but rather in doing something that no one has done before and that will add to the knowledge appreciated and acknowledged by the scientific community.

Scientists must be original if they are to be recognized for their work. The problem of originality is pressing not only because scientists want recognition, but because there are at times other scientists working on similar or closely related research who might arrive earlier at the same results. Scientists working in areas where others are working or knowing that others could be working on the same problem try to arrive, as Reif (1962) puts it, "fustest and mostest."

One result is competition for priority of discovery. The general limits of current knowledge and the accepted paradigm (Kuhn, 1962) help define what should be looked for, and the current research technology determines how or whether it can be done. The scientific community recognizes research that contributes to the solutions of the important scientific problems and its recognition is the principal reward of the discoverer. Other rewards such as promotion, fame, prizes, flow from such recognition.

Unlike the artist or author who has an infinite variety of paintings to paint, sculptures to sculpt, or novels to write, the scientist has only one world to discover. At any one point in history the frontier in a research area is essentially the same for all scientists doing research in that area. Scientists trained in the art of research, whether they are men or women, black or white, Chinese, Russian, American, or French, are going to make the discoveries next to be arrived at on the research frontier. The Mona Lisa, David, or Saint Peter's may never have existed without the unique inspiration of specific artists, but if Rutherford —as great as he was—had not shown that the atom could be split, someone else would have soon done it anyway. Einstein's name is a household word, and contemporary science acknowledges his contribution, but his reputation is what it is more by the fact that he developed his theories of relativity before others did than by the fact that only he was able to develop them. We would not have the Fifth Symphony without Beethoven, but we would have had relativity without Einstein.

Originality in science thus includes a time dimension that is especially significant. In art, the chances that two creative artists will produce exactly the same sculpture or painting are extremely low, but in a competitive field of science the chances are very high that two or more creative scientists will make simultaneous discoveries. Each artist may be creative and produce a work with great originality and be recognized for it. While each scientist may be creative in the research process, the one who completes and announces his results first will receive more recognition than the other scientist whose work may have been just as creative and *almost* as original.

The problem of originality in science results from the incompatibility between the requirements of scientific research on the one hand and the human requirements for recognition on the other hand. Requirements for the fastest scientific progress ideally involve behavior that puts a higher priority on achieving the community goal of progress than on the personal goal or recognition. The goal of scientific research, Robert Merton (1957:550–61) argues, is to extend "certified knowledge." This involves a set of conditions necessary to achieve the goal. He says that as a social institution science has four norms or rules. Each norm is not unique to science, but together they are peculiar to the ethos of modern science.

First, *organized skepticism* requires that all new knowledge must stand up to the same scrutiny, regardless of its source, before it can become a part of the accepted body of certified knowledge.

Second, *universalism* requires that age, sex, race, creed, or any special consideration should not enter a decision about the acceptance or rejection of scientific information. The relevant criteria in accepting or rejecting information establishing a new "truth" or revising an old "truth" are the logical structure and the quality of the data.

Third, *communism* (or communality) requires that once a particular piece of information has been thought up (or discovered) and made public, the originator of the new knowledge has no future intellectual claims to it and therefore all scientists are free to use it in future work. Indeed, there is a presumption that scientific information will not be kept secret but published within a reasonable time.

Fourth, *disinterestedness* requires scientists to be motivated to extend knowledge, not to seek personal gain. Scientists must have money to live on as does everyone else, and some do make large salaries, but material gain is not the main reason for pursuing research. Even if a scientist pursues materialistic goals, his research must not reveal his real motives. Disinterestedness also implies that the "truth" is to be accepted regardless of where it may lead or the implications therefrom.

These norms were not derived from the opinions of scientists, and all scholars would not necessarily agree that these are the norms of science. They are compatible, however, with the progress of science. Making the assumption that these are the norms of science allows us to observe the actions and processes in the scientific community and compare them with this model of science.

An exact fit between the operation of the scientific community and the normative model probably never obtains because scientific research is conducted by humans in a social environment. James Watson (1969:ix) writes in *The Double Helix* that the process of science seldom takes place in the straightforward and logical manner that it appears to do to those outside of science, but rather, ". . . its steps forward (and sometimes backward) are often very human events in which personalities and cultural conditions play major roles."

A major goal of this book is to show how the internal operation of science results in dilemmas for scientists because the norms prescribe the acceptable ways for researchers, as scientists, to act, but scientists are human and want to advance their careers.

The universalism norm is a problem for science when the educational backgrounds and positions at higher prestige universities affect the opportunities for scientists to be productive and recognized.[2] Disinterestedness presents a problem for scientists; while discovery is supposed to be for the benefit of scientific progress, scientists want recognition for their original work. The desire for recognition, in turn, presents problems for the norm that demands that scientists share the findings of their research. It is not so much a question of whether or not a scientist will adhere to this norm as it is of when he will inform the scientific community of his results. If a scientist divulges his results at the wrong time, he might aid his competitors to make a discovery and lose his rightfully earned recognition; if he does not share his results, he may be violating a norm of the scientific community.

British high energy physicists are the subject matter of my book. Sometimes also called elementary particle physics or fundamental particle physics, high energy physics was the most exciting physics research frontier during the 1960s and during the 1970s is expected to reveal new insights into the nature of matter. Because high energy physics has a large number of researchers, enormous funds, and a rapid rate of discovery in pushing back the frontiers, originality and competition are significant problems for scientists in this field.

Experimental high energy physicists use machines popularly known as "atom smashers" that accelerate particles (protons or electrons) almost to the speed of light, causing them to collide with other particles while the physicists observe the "interactions." Theoretical scientists compare data with phenomenological models, build theoretical models, or try to explain the physical process with abstract mathematical theories.

I was interested in studying British scientists because the organization of education and science in Britain is so different from that in the United States. In Britain, whether one attends a university greatly depends on the kind of secondary school one attends, a choice that is made when a student is about eleven years old. This process differs from university selection in the United States where a majority of students may decide whether they want a college education at about age seventeen. If admission is denied at a student's first choice, there are many other universities where one may apply (see Trow, 1962).

American students are geographically mobile at the

stage of graduate education. Many undergraduate colleges do not award Ph.D. degrees, and there are taboos against inbreeding in education that all but require students to take their advanced degrees from an institution other than their undergraduate college. The requirements for the Ph.D. degree and admission to Ph.D. programs vary among universities, and some students shop around. And not the least important reason for attending a specific graduate school or university is the financial aid available and the prestige of that particular university. British students, in contrast, usually remain at their undergraduate university if they take Ph.D.s. Furthermore, a prerequisite for doing research leading to a Ph.D. is usually a first-class honors undergraduate degree and almost always at least a second-class honors degree, whereas in the United States many institutions do not have such high standards for admission to graduate programs.

University finance in Britain is centralized, and the universities comprise a rather homogeneous system unlike the great diversity in the United States. There are few universities compared to the number in the United States, and this means there are relatively few opportunities for mobility between universities.

The British government distributes research funds through a centralized system, much different from the decentralized method used in the United States. The various British research councils are responsible for specific areas of scientific research. In the United States, research support for a variety of disciplines may be obtained from many sources such as the National Science Foundation, the Department of Defense, the Office of Naval Research, the Atomic Energy Commission, the National Institutes of Health, and others.

High energy physicists are among the elite of all British scientists, and they enjoy a certain social position. They spend the lion's share of funds that the government allocates for basic research in universities. In the 1967–68 fiscal year, the HEP[3] community spent over sixteen million pounds, more than one-fifth of all government funds allocated to scientific research in the universities.

Being favored economically by the government provides part of their high status among other scientists. Also important, however, is the fact that HEP work on perhaps the most esoteric of all research, a factor that contributes to the allocation of social status in scientific circles. Basic research has a higher status than applied research. The more basic the research, the greater the social status awarded to the researcher is a proposition that has not been systematically tested, but there seems little doubt about its validity among observers both in and out of science. As C. P. Snow (1959:32) writes:

> Pure scientists have by and large been dim-witted about engineers and applied science. They couldn't get interested. They wouldn't recognize that many of the problems were as intellectually exacting as pure problems, and that many of the solutions were as satisfying and beautiful. Their instinct—perhaps sharpened in this country, by the passion to find a new snobbism whenever possible, and to invent one if it doesn't exist—was to take it for granted that applied science was an occupation for second-rate minds. I say this more sharply because thirty years ago I took precisely that line myself. The climate of thought of young research workers in Cambridge then was not to our credit. We prided ourselves that the science we were doing could not, in any conceivable circumstances, have any practical use. *The more firmly one could make that claim, the more superior one felt.*[4]

Whether all scientists feel this way is not the question, but status differences based on that criterion are an interesting sociological phenomenon.

The high energy physics scientific community in Great Britain at the time I did this research was comprised of approximately 220 scientists, located at the following universities and laboratories: Birmingham, Bristol, Cambridge, Durham, Edinburgh, Exeter, Glasgow, Hull, Imperial College of Science and Technology (London

University), Kent, Lancaster, Liverpool, Manchester, Oxford, Queen Mary College (London University), Sheffield, Southampton, Sussex, University College, London (London University), Westfield College (London University), Atomic Energy Research Establishment (Harwell), Daresbury Nuclear Physics Laboratory (Daresbury) and the Rutherford High Energy Laboratory (Didcot).

I obtained the names of university scientists from *Scientific Research in British Colleges and Universities* (Department of Education and Science and the British Council, 1967). My sample included all Ph.D. scientists with some teaching duties. This eliminated most scientists who had just completed their theses and were on a temporary postdoctoral grant to complete some work in progress. They were excluded because they were not really in their first position, and as subsequent conversations substantiated, they were not yet really a part of the active scientific community.

At the laboratories, my sample included Ph.D. scientists whose research was mainly in elementary particles.[5] Scientists who operate accelerators and other machines were not included because they are more akin to engineers than scientists and do not experience the same kinds of influence in the course of their careers.

I conducted 159 interviews at the universities (including three self-administered interviews) and 44 interviews at the laboratories. Eleven scientists were unavailable because of travel and six refused to be interviewed although there were several others in the same department who cooperated. Out of a defined population of 220 scientists, 203 (92 percent) provided information for the study, and most of these (180) were tape-recorded (see Gaston, 1969).

I will discuss various aspects of the problem of originality and competition, including (1) the scientists' social, educational, and professional background; (2) the kinds of research conducted; (3) contributions to the scientific community; (4) recognition received from the scientific community; (5) attitudes toward research; (6) the prevalence and severity of competition; and (7) communicative behavior. Wherever possible, I will compare my data with those from studies of American scientists[6] to show that the somewhat different social and cultural conditions in Britain have an effect on the social system of science.

The Organization of Basic Research

2

In Britain, as in the United States, private industry, foundations, and the government are the three main sources of research funds. Private industry finances a small portion of university research, most of which is in the applied area (Hiscocks, 1959). Sponsored or foundation research is mainly devoted to social or educational research as in the United States at present. This chapter describes the organization as it was through 1972, but may not be the way it will be organized in the future.[1]

The main source of scientific research funds is the central government. In 1964–65, 756.6 million pounds were spent on research. Of this the government supplied 56.4 percent of the research funds, but government establishments carried out research amounting only to 25.4 percent of total funds expended. Universities furnished only 0.2 percent of the funds but spent 7.4 percent of the total. Private industry provided about 43 percent but spent about 67 percent (Council for Scientific Policy, 1966:18).

Research in the fiscal year 1964–65 conducted in the universities and intramural programs of the research councils amounted to 11.1 percent of the total research funds while other kinds of research and development amounted to 88.9 percent. All research conducted in universities is not pure research, but the applied research conducted in universities probably costs about the same as the pure research conducted in the defense and civil sectors. Therefore, when the interest is on the operation of the social system of basic science, the discussion is about scientists who are involved in university and research council research.

The central government supports both basic and applied research. Until 1965 one government department, the Department of Scientific and Industrial Research, had the major responsibility for much of the research financing. The Atomic Energy Authority, headed by a director directly responsible to the prime minister, and defense research controlled by the Minister of Defense were separate organizations, but nonmilitary government-sponsored research in industry and universities was administered by the Department of Scientific and Industrial Research. Since the enactment of the Science and Technology Act of 1965, the responsibility for industrial research has been placed in the Ministry of Technology (Mintech), and the responsibility for basic research has been placed under the Department of Education and Science. In dealing with scientific communities whose goals are to extend the frontiers of fundamental knowledge the Department of Education and Science is the department with major responsibility.

A Secretary of State heads the Department of Education and Science and makes his own budget, in consultation with other cabinet members. He knows that when the budget is prepared and agreed upon by the other ministers it will be approved by Parliament—quite unlike, for comparison, the director of the National Science Foundation in the United States who may not know until the fiscal year has begun, and sometimes even after it has begun, what funds will be available. The Secretary of State has a group of scientific advisers, the Council for Scientific Policy, from outside his department to guide him in scientific matters.

The Council for Scientific Policy is comprised of fourteen scientists who have positions either in universities, research institutes, or industrial organizations. Their advisory position is related to suggestions on general policy rather than to specific decisions. The council's first report (1966:2–3) to the Secretary of State for Science and Education stated:

Our terms of reference are to advise you in the exercise of your responsibilities for civil science policy. The range of problems has therefore become a narrower but much more intensive one [than when the council was the Advisory Council for Scientific Policy]; instead of advising on a very wide range of problems arising largely *ad hoc* and concerning frequently the organisations and spending powers of other Ministers, we have now to advise you in your statutory responsibility for determining the overall pattern of the resources of the Research Councils.

Because much of the advice on policy is related to university research, the University Grants Committee is an important element in the making of policy regarding scientific research, though its influence is different from that of the Council for Scientific Policy.

The University Grants Committee is the national funding agency for all British universities, and like the Council for Scientific Policy, the members are appointed by the Secretary of State for Education and Science. There are over twenty members of this committee, and each is either a university staff member or has had some experience in university affairs. The task of the committee is to distribute its quinquennial grant from the Exchecquer. It does this through distributing block grants to individual universities, which are then relatively free to distribute them as they see fit. Certain guidelines are expressed such as requests that some programs be started, consolidated, or phased out, but on the whole (with the exception of capital expenditures that are earmarked) each university has a great deal of freedom in making its own decisions.[2]

The University Grants Committee is involved in scientific research since all university staff members' salaries are paid from these block grants. Technicians' salaries, supplies, and other necessities for research (not funded from a research council or other grant) are paid from departmental funds that are budgeted by the university administration. Not all research salaries are paid out of University Grants Committee funds since some research grants include research assistants and postdoctoral stipends, but all regular university positions are paid through the committee. Since the University Grants Committee and the Council for Scientific Policy are complementary in many respects, the chairman of the UGC sits as assessor on many of the committees and councils that are involved with research funding of university-based scientific research.

In summary, the Department of Education and Science has the main responsibility for basic research in Great Britain. The Council for Scientific Policy advises the department on overall policy matters—which areas should be strengthened, which areas weakened, and so on. The UGC is responsible for distributing block grants to universities, but it is not directly responsible to the department for its decisions. Cooperation for a centralized national effort is the norm, however, and contact between groups is maintained. For example, unless the UGC is willing to approve additional staff positions in some research area, the appropriate research council will not approve research applications from a given university.

The research councils are the main source of research grants. Within the Department of Education and Science, there are four councils: the Agricultural Research Council, the Medical Research Council, the National Environment Research Council, and the Science Research Council. The British Museum (Natural History) is also governed by the department. The Social Science Research Council is in the department, but the Council for Scientific Policy does not advise on the SSRC's research goals. In the year 1967–

68, according to the Council for Scientific Policy (1967:39), the councils were estimated to spend 72,607 pounds ($2.80 to the pound). The amount to each was: agricultural—16.5 percent; medical—19.6 percent; natural environment—10.5 percent; and the large one, science—50.4 percent. (Other expenses consumed 3.0 percent.)

The research councils operate similarly to the United States National Science Foundation. They process research applications, grant fellowships, and sponsor training programs. One big difference, however, is that the research councils operate their own laboratories for in-house research and for the benefit of university scientists. The National Institutes of Health in the United States do maintain laboratories, but the Atomic Energy Commission contracts to university consortia the operation of its major laboratories. In Britain a national scientific policy is followed since the research councils are in almost complete control of basic research activity. That centralization compares very differently with the situation in the United States where the Department of Defense as well as the Atomic Energy Commission, the Office of Naval Research, or the National Science Foundation may sponsor physics research.[3]

The content of the agricultural and the medical councils' research interest is obvious. The National Environment Research Council is concerned with oceanography and geology. The largest spender of the research councils, the Science Research Council, is responsible for (1) astronomy, (2) biology, (3) chemistry, (4) mathematics and computing science, (5) engineering, (6) solid state and other physics, and (7) nuclear physics.

There are sixteen members of the Science Research Council who are either university professors or Ph.D. scientists. The work of the council is divided among various boards or committees. Scientists who are not members of the council are asked to serve on these committees with council members to widen the representation of the scientific community to the council.

The special interest in the Science Research Council stems from its support of high energy physics. For the year ending 31 March 1968, the Science Research Council spent 37,855,361 pounds on research under its province. Excluding the University Grants Committee's indirect support of high energy physics research through salaries, the Science Research Council itself spent at least 16.3 million pounds on HEP for the year ended 31 March 1968, an amount slightly over 43 percent of the total Science Research Council budget and about 22 percent of the total expenditure of all the various research councils. The fact that the number of scientists who spend this proportion does not approximate their proportion of all scientists in Great Britain underscores the expensive characteristics of this Big Science and emphasizes the relative value that the scientific community (through the Council for Scientific Policy and the Science Research Council) puts on HEP as a research priority.[4]

The Science Research Council operates two HEP laboratories. The Rutherford High Energy Laboratory has a 7 BeV (billion electron volt) proton accelerator, and the Daresbury Nuclear Physics Laboratory has a 4 BeV electron accelerator. The Rutherford High Energy Laboratory is located about fifty miles from London (and fifteen miles from Oxford) and primarily serves the universities located in the south of England. The Daresbury Nuclear Physics Laboratory is located about fifteen miles from Liverpool and primarily serves the Scottish universities and those universities in the north of England.[5] The Science Research Council provides about 20 percent of the operating budget of the European Organization for Nuclear Research (CERN), an international laboratory at Geneva. CERN has a 28 BeV proton accelerator and the primary British usage of the facility has been to obtain bubble chamber film for analysis back at the scientists' own universities.

To complete the introduction to the various components of the organization of basic scientific research in

Great Britain, a brief discussion of the universities is necessary. Before the middle 1950s there were twenty-four universities in Great Britain. By the time of the Robbins Report in 1963 there were thirty-one universities. The increase of seven universities resulted from one university with two colleges becoming two separate universities, another college receiving university status, and the founding of four new universities. The main recommendations of the Robbins Report, which was accepted by the government, were to expand greatly the number of student places in universities and to increase the number of universities. The increase in the number of universities came about through both the creation of totally new universities and the "promotion" of some colleges of advanced technology to university status.

There are thirty-six institutions in England receiving support from the University Grants Committee. That figure includes London University as one institution although it has many constituent colleges. Included in the thirty-six institutions are two business schools (London Business School and Manchester Business School) and one institute of technology (Manchester Institute of Science and Technology). Wales has one university with several constituent colleges located in different towns. Scotland has eight universities, four of these dating back to the time when Oxford and Cambridge were the only two English universities. In total then, forty-two universities, two business schools, and one institute of science and technology for the most part constitute the British system of higher education. This figure excludes, of course, the teacher-training colleges as well as technical and other schools not enjoying university status.

For the 1967–68 academic year, the University Grants Committee distributed 150,790,000 pounds to the various institutions with individual university allotments ranging from 336,000 pounds for Stirling University, Scotland, to 32,159,200 pounds for London University.[6] It will be recalled that basic research funds from the Science Research Council for HEP during the same academic year totaled about 16.3 million pounds. As another comparison, therefore, HEP cost approximately 10 percent of the total for operating the whole university system.

FIGURE 1

ORGANIZATIONAL COMPONENTS OF THE HEP COMMUNITY

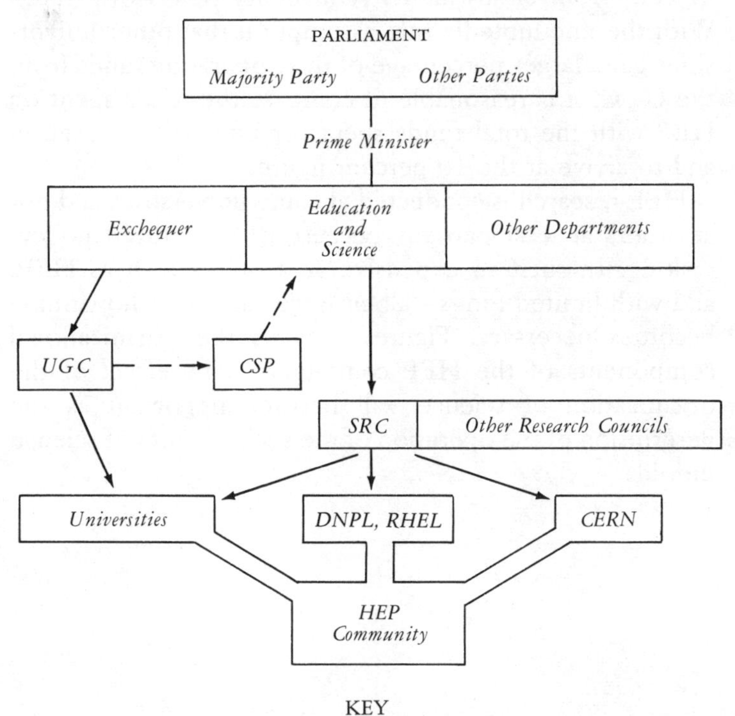

KEY

CSP—Council for Scientific Policy
UGC—University Grants Committee
SRC—Science Research Council
CERN—European Center for Nuclear Research
DNPL—Daresbury Nuclear Physics Laboratory
RHEL—Rutherford High Energy Laboratory
HEP—High Energy Physics

Funds: ———————————

Influence: – – – – – – – – – – –

The University Grants Committee funds do not comprise 100 percent of the university operating funds. Oxford and Cambridge, however, are the only universities with endowments of any significance. Oxford's income statement for the 1966–67 academic year, the year before the figures reported above, showed that of the 6,327,352 income (excluding research contract income), 84.3 percent came from the University Grants Committee and only 7.1 percent from endowments (University of Oxford, n.d.). With the undoubtedly safe assumption that other universities get a larger percentage of their operating funds from the UGC, it is reasonable to compare the funds spent on HEP with the total funds spent for university education and to arrive at the 10 percent figure.[7]

HEP research is conducted at some universities and not at others at least partially because of centralized policy-making. In such an expensive area of research as HEP, and with limited funds—albeit large sums—such planning becomes necessary. Figure 1 shows the organizational components of the HEP community. The effect of the organization of science will become important as the description of the operation of the social system of science unfolds.

Research in High Energy Physics 3

Physics is the study of matter, and the study of elementary particles is the study of the smallest "bits" of matter that are presently known to exist.[1] Put more explicitly by the High Energy Physics Advisory Panel (1968:11), "High-energy physics is one of the main fronts of science. . . . It tries to establish the fundamental laws of physics which are at the base of all that we know about matter." According to the panel, research looks "for the laws governing the four fundamental interactions—nuclear [strong], electromagnetic, weak, and gravitational—with the final aim of unifying these interactions by finding some common origin." In addition, ". . . apart from seeking an understanding of 'interactions' between particles, high-energy physics seeks to find reasons for the existence of particles themselves. Why is matter made of nucleons and electrons?"

Physics is among, if not the most differentiated of all scientific disciplines in terms of subdiscipline boundaries. Many of the research specialties of physics such as optics, mechanics, and fluids have been present throughout much of the history of physics. Nuclear physics and its daughter elementary particle physics are essentially physics of the twentieth century. Nuclear physics was born when

Rutherford of the Cavendish Laboratory in Cambridge showed that the atom was a divisible piece of matter in 1919. In the twenty years after that, nuclear physics was frontier physics.

The early particle accelerators were constructed to study the results of protons causing the disintegration of light nuclei. Two British scientists, John Cockroft and Ernest Walton, were the first successfully to observe this, and they did so at the Cavendish Laboratory in 1932. M. Stanley Livingston, an American, remarks that "I like to consider this as the first significant date in accelerator history and the practical start of experimental nuclear physics."[2] Once this research started, it generated its own momentum.

The early history of nuclear physics is marked by the building of successive generations of accelerators. Improvements were made as each accelerator formed a springboard to the next. Machine designs were changed as the principles used in early ones could not be used in machines of higher and higher energies because the design would have called for unwieldy parameters (this progress has continued until the present day).

Then the war came, and basic research as previously organized gave way to defense research. Although normal research and the open communication of results were slowed down during the war, in the end the war was probably good for the progress of nuclear physics. The potentials of nuclear physics were recognized, and the stage was set for the introduction of really Big Science, a type of science that requires the social organization of research in large laboratories similar to those organizations created after scientists were mobilized in the United States during the war.[3] This is not to suggest that organizing a large number of scientists in laboratories, as is required by Big Science, is good; it is to suggest that the nature of research in nuclear physics requires large-scale organization and therefore the war experience probably assisted this unforeseen but inevitable development.

After the war there were large numbers of scientists ready to return to basic research. Because nuclear physics had contributed so much to the war effort, funds for that type of research were readily available. In the United States, where both British and American scientists had worked closely on developing the bomb, the government's Manhattan District set a precedent for government support and this came to be the norm for research in nuclear physics. When an outstanding wartime scientist, E. O. Lawrence of Berkeley, suggested that the Manhattan District support the construction of an accelerator started before the war at Berkeley, General Leslie Groves of the district agreed to supply $170,000 for that purpose. By the end of 1946, two accelerators were in operation at Berkeley (Swatez, 1968:58):

> These post-war developments just prior to the establishment of the Atomic Energy Commission were important for the [Berkeley] Radiation Laboratory's future place in the scientific community and in the University. They set the precedent for continued support by the AEC and signified the orientation of the Laboratory as a national laboratory in practice if not quite formally.

By the end of 1946 Manhattan District contracts and installations were transferred to the newly established Atomic Energy Commission. By 1952 accelerators were completed with energies high enough to provide a source of particles that would produce interactions of great interest to the physicists.

Subnuclear particles, that is, elementary particles, began to be studied using accelerators only after World War II. Prior to that time the minimum threshold energy (about 200 million electron volts) necessary for a proton to collide with an atomic nucleus and dislodge other elementary particles from the nucleus was not available.

Research on elementary particles can be carried out without accelerators. In fact, accelerators produce particles of much less energy than are found in nature. The "natural" particles are called cosmic rays, but the origin(s) of cosmic rays is not fully understood by scientists. Although some come from the sun, others come from space much farther away than the sun. As the cosmic rays, mostly protons, come into the earth's atmosphere, they interact with atoms and produce new particles that continue toward the earth. It is difficult to get a large number of interesting "events" (particles interacting with other particles) of a given type with well-known parameters at ground level using cosmic rays, since at that level muons are the predominant particle, but cosmic ray physicists do set up some apparatus for observing the particles at ground level. They also launch balloons with detection devices aboard to observe these particles as well as place detection equipment in rockets.[4] Scientists prefer more control over their experiments than cosmic rays allow, so accelerators have been favored over cosmic rays since experiments in the laboratory (at a much lower energy, of course) can be done with a greater precision and larger statistical advantage.[5] Some physicists prefer to study cosmic rays and therefore provide the distinctions between high energy physicists (accelerator users) and cosmic ray physicists, who really use higher energies.

In the quest for knowledge about particles, larger and larger accelerators have been designed and built. Although universities were the original locations of these accelerators built with governmental support, it soon became necessary for accelerators to be built for several users. Consequently, the development of national laboratories is now the mode in all countries doing elementary particle physics research using high energy accelerators.

The geographical distribution of high energy physics research is highly correlated with the wealth of nations. Consequently, there are three main regions carrying on extensive research activities—the United States (and, to a lesser extent, Canada), western Europe, and the Soviet bloc. Research is conducted in Japan, India, and China, but the extent of the effort in China is uncertain. Of course the problem of how to measure "how much" is introduced since number of scientists, scale of research expenditures, or number of papers produced may give three different but "valid" estimates of the extent of research activity.

As to the raw number of scientists, the question of nationality or nation of employment enters into the problem of assessment. Although no really sound data or census exist, a study by the American Institute of Physics of theoretical scientists gives some clue (Libbey and Zaltman, 1967:24–25). For example, whereas 30.4 percent (out of 977 respondents) were American nationals, 42.3 percent were employed in the United States. Conversely, whereas 6.0 percent were Indian nationals, only 2.4 percent were employed in India.[6] Taking these figures, one finds that in theoretical high energy physics, 32.3 percent are American and Canadian, 35.0 percent are western European, and the remainder are scattered throughout many countries. The lack of information about the Soviet Union and China means that any estimate is subject to considerable variation; however, it would not be grossly inaccurate to suggest that the three major areas have about three-fourths of the theoretical high energy physicists and another one-fourth are scattered around the world. Experimental work is more concentrated because it is only done where accelerators are available whereas theoretical work may be done in the most extreme backwaters (though the quality of research in these locations may be affected).[7]

On the experimental side there is again inadequate data about the number of researchers from the Communist countries. In terms of accelerator capabilities, however, it is easy to see that western Europe and the United States are rather comparable; and although the Soviet Union has fewer accelerators, it now has in operation an accelerator with over twice the energy of any other accelerator. The distribution of the three largest particle accelerators is as follows:

USSR	76 BeV (Serpukhov)
United States	30 BeV (Brookhaven)
Western Europe	28 BeV (CERN)

The United States has built a machine that is not yet fully operable with original energy of 200 BeV, potentially convertible to 400–500 BeV. That machine is at the National Accelerator Laboratory in Illinois. CERN has plans for a 300 BeV accelerator.

Experimental High Energy Physics

In experimental high energy physics, particles (protons or electrons) are accelerated to almost the speed of light and are then caused to collide with stationary targets with the purpose of producing subatomic elementary particles of matter.[8] This enables a study to be made of their properties and interactions. Because particles are not visible to the eye, various methods have been developed for "detecting" particle interactions. A very early device was a cloud chamber that allowed the particles to be photographed as they traveled through a box filled with a gas and liquid vapor so that when the gas was cooled the vapor was supercooled and an electrically charged particle caused tracks visible in the vapor. Nuclear emulsion targets have also been used for detection of particles. More advanced techniques include counters, spark chambers, and bubble chambers, and this instrumentation is the essence of Big Science.

Bubble chambers[9] were originally small "boxes" holding a liquid target, but as everything else has grown more sophisticated in high energy physics, bubble chambers have also. They are so expensive to build that they are not maintained for one group of scientists but operate as a service to many groups.[10] Groups of scientists "order" a certain number of pictures specifying at what energy they want a certain beam of particles to be sent into a bubble chamber. They also specify the liquid target—the most common target being hydrogen. As the accelerator pulses and particles are directed into the chamber, the chamber releases pressure on the liquid that allows bubbles to form around the nucleation centers provided by the ionization of the liquid molecules produced by the passing charged particles. The "events," or interactions, are photographed from various angles and the pictures are then analyzed, reconstructing a three-dimensional model of any "event" of interest that occurred.

Analysis of the pictures is carried out on a large scale. Pictures must be scanned for potentially interesting events, and then these pictures are measured and recorded for computer analysis. Large numbers of human scanners are employed, and a large amount of computer time is required. It is not at all unusual for a team of bubble chamber scientists to have a computer system as large as most universities have for all of their users.

Depending upon the research problem, bubble chambers may be placed in magnetic fields causing charged particles to be deflected in various directions so that velocity, mass, and angles may be analyzed. Freon, deuterium, propane, xenon, and neon are also used as targets in bubble chambers when particle interactions on neutron or heavy nuclear targets are to be observed, but because of the more complex nature of their nuclei the results are more difficult to analyze.

Spark chambers are essentially of three types: visual, sonic, and wire. The visual spark chambers record events by photographing what happens when particles interact, and these film are treated similarly to bubble chamber photographs, although there are relatively fewer pictures involved since the spark chambers are activated only if a signal has been given indicating that a potentially interesting event has occurred. Sonic spark chambers record sounds through microphones strategically placed to record distances. The sounds are made as high voltage sparks occur on the path of electrically charged particles through arrays of charged metal plates. The time of arrival of these sound data are recorded magnetically and fed into computers, which in some cases are operated on-line directly to the experiment. Wire chambers are used in a manner similar to sonic chambers, but it is the position of the wires that give the coordinates of the particle rather than the time taken by the spark shock wave to reach the microphones in the sonic chamber. Spark chambers are used in conjunction with counters in many experiments.

The generic name "counter" applies to a variety of devices whose function is simply to count the number of particles that come in contact with its sensitive areas. Counters usually work in conjunction with other electronic detection devices so that a reconstruction of a particular event is possible.

Bubble chamber experiments are "dirty" experiments —to use the physicist's vernacular—meaning that scientists capture on film everything that happens in the chamber, and all these pictures must be viewed for possible interesting results. Thus, research really begins after the film is developed. Having everything that happened during an experiment has an advantage: research may occur in which occasional findings turn out to be more interesting than the analysis previously planned.

Counters and spark chambers, in contrast to bubble chambers, produce "clean" experiments in which the exact research question is all that may be answered (if, indeed, it can be). More planning is necessary, and some physicists like to say that counter research is the true physics while bubble chamber work is a problem in organization. This claim about bubble chamber research is true to some extent since a group of technicians operates the huge and complex bubble chambers while their counter brothers may participate not only in designing and building (and repairing!) their detection apparatus but actually set them up, test, and operate them for data-taking. Bubble chamber physicists (though some members of the team are present during data runs) do most of their work not on the accelerator floor but in their offices and in their group's computer centers. This is not to say that electronic experimentalists are always "in the shop." One experimentalist in commenting about bubble chamber specialists and their computing problems said:

> Don't forget the poor "counter" bloke who has just spent about six months running an experiment and then faces the next six months with 300,000 events (clean ones!) and lots of computer programming and sums to drag the physics out! Even counter experiments are not yet run on-line to the *Physical Review!*[11]

Theoretical High Energy Physics

Theoretical physics is strongly mathematical and even though all physicists must have a certain mathematical proficiency, theorists must have more mathematical aptitude than experimentalists. In fact, some physicists indicated that they became experimentalists partly because of their lack of mathematical ability.

According to theorists themselves, there are three types: phenomenologists, intermediate, and abstract. If this grouping is meaningful, and they say it is, there is considerable differentiation not only between theorists and experimentalists, but also within elementary particle theory itself.

These three types of theorists are located on a mathematical continuum. It is sometimes said that phenomenologists are frustrated experimentalists and that abstract theorists are frustrated mathematicians, because of the closeness these groups have to the others.

Phenomenologists use experimental data to construct and test models of how particles interact while abstract theorists are concerned with highly sophisticated mathematics.

Intermediate theorists, who say they are neither of the other two types, usually indicate a leaning toward phenomenology. When asked why, they answer that they want to feel that their research is useful. The status of grand theory in elementary particles is very low since an adequate theory does not exist (or if it does exist, it is not recognized). An intermediate theorist said that he would not like to consider himself on the abstract side of theory. When asked why, he replied:

I think the justification for high energy theory is accounting for facts. We theorists don't do pure mathematics well enough to be called mathematicians. So, really abstract research—I would say—is not terribly valuable. If it's pure mathematical research, it should presumably be of interest to pure mathematicians, and it's not, in general. If it's not pure mathematics, it should be attempting to account for something.

Abstract theorists are concerned with axiomatic approaches and are, in the main, not concerned with the connection between physical reality and their theory, or at least they are so accused. Elegance and beauty of presentation are important, however.

Based on my personal observations in the process of associating with HEP for several months, it appeared that a certain status differentiation exists between theorists and experimentalists. For example, an experimentalist said that

Most experiments are just acquiring extra data, which is then put into the melting pot, and we're happy if we just get some numbers and cross-sections of branching ratios. We don't bother too much about the interpretations of it.

This statement in itself does not substantiate any status differences, but there is other evidence. A theorist told Hagstrom (1965:250):

To a large extent I think [the interpretation of experiments] should also be left to the theorists. Experimentalists are apt to bungle results and present them in a form that is incomprehensible to the theorist—not only incomprehensible; sometimes they simply remove most of the information [Hagstrom's brackets].

In addition to experimentalist-theorist status differences, there also appeared to be differences within these types. The abstract theorist sees himself higher than the intermediate theorist, who sees himself higher than the phenomenologist, who sees himself higher than the experimentalist (who, after all, once his measurements are made is thought not to know what they mean). Bubble chamber experimentalists seem to have higher status than counter physicists. Many times people alluded to the possibility that social status determined the floor where one's laboratory was located, and that bubble chamber physicists were always on a higher floor. Whether or not these remarks were defense mechanisms is difficult to determine because bubble chamber specialists sometimes pride themselves for "having to fly off to CERN" at regular intervals, while some scientists made statements that indicate a different attitude toward counter/spark chamber specialists. One bubble chamber specialist said:

One of the troubles with bubble chamber physicists is that there are not very many good physicists. The very good physicist [experimentalist] seems to go into other fields—the counter experiments, for example. Those people usually think of experiments they would like to do and then go and do them whereas people that work in bubble chamber film analysis are more organizational men, I think. Bubble chamber film analysis is a question of organization and teamwork, and usually your capacity as a physicist doesn't make or break the experiment. What's of interest is the results you get—not necessarily your interpretation of them. Once you've got the film, you're guaranteed to get the results—this is not the case in counter work.

The Reward System

4

The number of papers scientists publish in their lifetime varies enormously. Price (1963:45–46) calculates that 3.54 papers is the average number, but "Scientific papers do not rain from heaven so that they are distributed by chance; on the contrary, up to a point, the more you have, the easier it seems to be to get the next."

What difference does it make that scientific productivity varies? While the production of scientific papers is only one indication of a scientist's contributions to his professional community, it is to a large extent a measure of how much he is in fact a scientist. A person may be connected to science as a teacher or an administrator, but if he does not publish, he is not a scientist.

The socialization process, to a large extent based on apprenticeships and following examples, includes internalizing the notion that the purpose of science is to create knowledge through research. Research results are formally disseminated through scientific papers. If production of scientific papers is an indication of a scientist's worth to the community, and because productivity does vary, then some scientists are apparently worth more than others.

Is it necessary to reward a scientist for his contributions? The answer seems to be an overwhelming yes. Merton

(1962; 1965) has demonstrated without question that one of the main reasons why scientists do research is to obtain recognition from their peers. A further condition can be specified. Scientists do not necessarily become scientists to do research for recognition, but they continue to do research in the hope that recognition will come their way. Failure to obtain recognition over a period of time might prevent some scientists from continuing their research, and while data are limited for this speculation, some probably continue to do research long past the time when they might have been recognized. The fact that many eventually do receive recognition is probably sufficient impetus to keep up the work.[1]

The institution of science rewards scientists with recognition for their work. Science would not progress if scientists were not retained in the community, thus there must be some correspondence between the contributions a scientist makes and the rewards he receives for his contributions. The nature of the reward system is the specific way in which scientists are accorded the recognition they merit as a result of their original research contributions.

The reward system in the United States operates in a diverse educational system with a highly visible prestige system.[2] There are several consequences of the prestige hierarchy for scientists' opportunities in being able to successfully engage in original work that will also stand the competition from other scientists.

Scientists trained at the more prestigious universities are exposed to a socialization process that helps them get a head start in becoming successful scientists. This is due partly to selective recruitment at the more prestigious universities and partly to the superior facilities and faculties available. Graduates from these universities are more productive than scientists trained at less prestigious universities, and although productivity is positively related to the prestige of the university in which a scientist may currently be affiliated, scientists trained at major universities are more productive regardless of their current affiliations (Crane, 1965; Hargens and Hagstrom, 1967).

Prestigious sponsors (thesis advisers) provide some advantage as well, and prestigious sponsors are more likely to be at the more prestigious universities. Cole and Cole (1968:400) report a product-moment correlation of .54 between prestige of scientists' highest awards and rank of department and .50 between number of awards and rank of department. Although productivity, according to Crane (1965), is related more to the prestige of scientists' graduate schools, recognition for productivity is related to the prestige of scientists' current affiliations. Quality of research may be a factor here since it was not controlled for in Crane's study, but Cole and Cole (1968) show that among eminent American physicists quality and quantity are strongly related.[3] The more prestigious universities recruit the best faculties and students whose presence reinforce each other. This cycle continues with only slight variation through time (Cartter, 1966).

Prestigious universities not only provide greater opportunities for scientists' productivity, they also provide greater opportunities to receive recognition for their work. This results from the fact that eminent scientists are located at the more prestigious universities and tend to control institutional mechanisms for rewarding deserving scientists. It is expected that rewards go to scientists who are known in some way by this community. Being students of these eminent scientists or students of their slightly less well known colleagues, therefore, provides differential access to the rewards of the community.

A study of British scientists must consider the problem of whether the social backgrounds and the institutional hierarchies affect scientists' opportunities to do original research and be rewarded because these factors apparently do affect to some extent opportunities for scientists in the United States. Additionally, a study of one particular research specialty must consider that, apart from the uniqueness of its content and work organization, there are social differences between the scientists in that specialty and the remainder of the scientific community. If the high energy physicists are very different from other British scientists, then the conclusions might be suspect on that account; if scientists have similar backgrounds, however, then differences between scientific specialties may be viewed more a matter of cognitive than social differences.

Social and Educational Origins

Social class, divided into five categories, was determined on the basis of fathers' occupation when the scientists were twelve years old, on the assumption that social class effects for education were probably established by that time.[4] There are 14 percent in Class I; 38 percent in Class II; 33 percent in Class III; 13 percent in Class IV; and 3 percent in Class V (the total of 101 percent results from rounding). The social class origins of HEP are very similar to a large sample of British academics (all disciplines included) and even more similar to a sample of British scientists that excluded technological scientists (Halsey and Trow, 1971:216, 413). The largest percentage difference between the HEP and the sample of scientists—7 percent—occurs at Class IV and probably has no substantive significance. British academics in general are more likely than scientists in particular to originate from Class I because all disciplines are included in academics and scientists generally are from lower social origins than classicists, for example.

The majority of HEP, 69 percent, attended a local grammar school for their secondary education. Compared to the 21 percent of British academics who attended public schools, a slightly smaller 16 percent of the HEP did (Halsey and Trow, 1971:216). This is probably because more British academics than HEP are Class I, and public schools draw students largely from the upper classes.

As a group, HEP attended the various universities for their undergraduate degrees in about the same proportions as all British academics. Using data from the Robbins Report (Committee on Higher Education, 1963; Appendix Three:36), the percentages of HEP (with all academics shown in parentheses) obtaining their undergraduate degrees from Oxbridge were 27 percent (31); London, 19 percent (22); "Redbrick," 25 percent (23); Wales, 2 percent (5); Scottish, 11 percent (13), and foreign, 17 percent (5). The main difference comes from the numbers obtaining their degrees from foreign universities.

According to the Robbins Report (Committee on Higher Education, 1963; Appendix Three:22), recent expansion of university staffs has resulted in a lowering of the overall staff qualifications. Of those teachers recruited before 1958, 61 percent (of 1779) had first-class honors degrees; and of those recruited in 1959, only 52 percent (of 485) had first-class degrees. The percentages of science academics with first-class degrees for each period were 67 (of 587) and 50 (of 189), respectively. The quality of HEP undergraduate degrees are considerably better than both academics in general and science academics. Of the HEP with classed degrees (which omits holders of foreign first degrees), 70 percent have first-class honors degrees; 28 percent have upper second-class degrees. Certainly the HEP are an educational elite.

Before students get their undergraduate degrees, many think they would like to "stay on" and do research. Hutchings (1967) reports that 63 percent of the physics undergraduates in his study of five universities wanted to get a Ph.D., with physics having the highest percentage. Only those getting the best or higher class degrees are asked by the department to do so, however. This contrasts with the American system in which application may be made to several universities with the idea of selecting one university from among several that may offer admission.

In Britain, not only must a student be intellectually capable of working on a research degree, the conditions in his department must be favorable for his retention. If research on a theoretical topic is the main interest and the student has shown a sufficient potential, it is a matter of deciding which staff member will assume responsibility for an additional student. A system of thesis advisers is not used as it is in the United States. If experimental research is the interest, then the various specialty groups that desire another research student discuss the possibility with the students. When asked about his choice of research, one experimentalist said: "I was very vague on what I wanted to do. I talked to the various group leaders about what was going on at the time, and that particular project sounded the most interesting, and I went for it."

If research groups are in competition for students, recruitment tactics may be used. A bubble chamber experimentalist recalled that his group likes to tell a potential student that he may "have" to fly over to CERN at Geneva a few times a year (and that, of course, his travel expenses will be paid).

Financial support for research students is provided by the Science Research Council. Departments receive funds for a certain number of studentships, and these tend to be awarded to students who took undergraduate degrees in the same university. Unlike the National Science Foundation and other fellowships in the United States, students cannot get national awards and go to any university they choose. Institutional mobility is thus reduced because of difficulty in obtaining an award from a department at another university. In fact, 132 out of the 199 scientists with Ph.D. degrees attended the same institution for both degrees.

In science disciplines there generally appears to be considerable direction given to students as to the choice of thesis problems.[5] For the HEP, 76 percent (151 out of 199 with Ph.D.s) indicated their problem was assigned;

18 percent said they chose their own; and 6 percent said it was both partly chosen and partly assigned.

As a group, HEP received higher class undergraduate degrees than British academics in general. This is not unexpected since it is frequently suggested that the best scientists are attracted to the research frontiers. Without comparable data from other specialties it is not possible to conclude that the HEP chose their research specialties or had them chosen for them more often than other scientists. But in most cases becoming a HEP did not seem to result from any deliberate decision based on specific alternatives. Rather, the decision on the part of the HEP appeared to be more a matter of chance. There should be some social explanation for this process, however. Many HEP were recruited before there was high energy physics, meaning that they began their research careers in nuclear structure or low energy physics. Nuclear structure, however, was the frontier physics at that earlier time. Therefore, the higher qualifications of the HEP and the appearance of chance decision probably results from the fact that the information was obtained from the HEP themselves rather than from the department heads and group leaders who throughout the years were also involved in the decisions. What appears to have been a chance decision to the HEP is very likely to have been a deliberate choosing of the "best" students on the part of the senior scientists whose groups they joined, or theoretical professors who were "able" to take on additional students.

The Prestige Hierarchy

Some indication of the existence of a prestige structure comes from the Robbins Report (Committee on Higher Education, 1963:8):

> We wish to see the removal of any designations or limitations that cause differentiation between institutions that are per-

forming similar functions. Distinctions based on adventitious grounds, whether historical or social, are wholly alien to the spirit that should inform higher education.

Ben-David and Zloczower (1962:68) examined the effects of an academic hierarchy in Britain on innovation and stated that "Universities, like so much else in that society, arranged themselves in a relatively neat hierarchy." This hierarchy includes the Oxbridge (Oxford and Cambridge) universities at the top with London University second and the other "provincial" (Redbrick) universities distributed below. The Scottish universities are not discussed by Ben-David and Zloczower.

There is evidence that the graduates of the ancient universities of Oxford and Cambridge have advantages over other university graduates. The Robbins Report (Committee on Higher Education, 1963:8) mentions that a goal for higher education ought to be equal academic rewards for equal performance. "We are well aware," it states,

> that there are limits to the realization of this principle, and that status accorded by the world to a degree from an institution of long standing and established reputation may well be higher than the status of a degree earned in an examination of comparable severity in an institution of more recent foundation.

Halsey and Trow (1971:213–24) give considerable evidence for the high position of Oxford and Cambridge as compared with other universities. Qualitatively, they quote (p. 213) Bruce Truscott's *Redbrick University:* "For to Oxbridge all the best people eventually gravitate, whereas to Redbrick no-one, if he can help it, ever comes at all. . . ." They quote from a 1955 article in *Encounter* by Edward Shils and state that his characterization has remained broadly valid to the present. One paragraph shows the general idea (Halsey and Trow, 1971:214–15):

> The modern British universities, which in scholarship and science take second place to none in the world, have—despite efforts of the University Grants Committee and many worthy men who have loved them—been belittled in their own eyes.

They have never had a place in that image of the right life which has evolved from the aristocratic, squirearchal, and higher official culture. To those who accept this image, modern universities are facts but not realities. They would not deny that Manchester, Liverpool, Birmingham, and the other urban universities actually exist and yet they do not easily admit them to their minds. Oxford and Cambridge are thought of spontaneously when universities are mentioned. If a young man, talking to an educated stranger, refers to his university studies, he is asked 'Oxford or Cambridge?' And if he says Aberystwyth or Nottingham, there is disappointment on the one side and embarrassment on the other. It has always been that way.

Halsey and Trow (1971:216–18) list three quantitative indicators of the reputation of Oxbridge universities: (1) Oxbridge dons (teachers) are more likely to come from professional, managerial, and white-collar classes; (2) Oxbridge dons are more likely to be educated at public schools; and, (3) over a third of the Fellows of the Royal Society in 1960 were affiliated with Oxbridge (a proportion that has been relatively stable even with a decline in the proportion of all teachers who are affiliated with Oxbridge), and over 60 percent of the Fellows of the British Academy in 1961–62 were affiliated with Oxbridge.

Oxbridge graduates are overrepresented in the professorial rank. Whereas 31 percent of all teachers surveyed by the Robbins Report (Committee on Higher Education, 1963; Appendix Three:38) were Oxbridge graduates, 41 percent of the professors were Oxbridge graduates. Because most professors are also department heads (Halsey and Trow, 1971:375), it can be assumed that Oxbridge graduates are also overrepresented as department heads, and they conclude (p. 225) that ". . . in general it is clear that any contact with Oxford or Cambridge improves the chances of reaching the highest academic rank elsewhere."

The question is not whether Oxford and Cambridge enjoy a high reputation. The question is, does their reputation guarantee any strong advantage *in the scientific community* to scientists affiliated with these universities? First it must be established whether scientists see the various research groups as having a pecking order. Then, any effects of this ranking must be considered.

To determine whether scientists believe a difference exists between the quality, and therefore the prestige, of various high energy research groups, I asked them to rate the groups located at various universities (excluding the laboratory groups) according to five categories: (1) not sufficient to provide doctoral training; (2) marginal; (3) adequate; (4) strong; and (5) distinguished. The question, adapted from Cartter (1966:127) and changing only a few words to conform to British terminology, was: Which of the terms best describe your judgment of the quality of the staff in high energy physics at each institution listed? Consider only the scholarly competence and achievements of the present staff. Limit the number of "distinguished" ratings to no more than two. Cartter limited the number of "distinguished" ratings to five, since he was usually rating more than fifty departments. Because there were about half that many departments being rated in this study, I arbitrarily limited the number to two (see appendix for details).

The university HEP were divided into three categories, according to their current affiliations: (1) high prestige, (2) middle prestige, and (3) low prestige (see table 1). An important question to be answered here is why some groups are believed to be more prestigious than others, i.e., why is their quality believed to be higher? One simple answer is because the quality of the staff may in fact be higher at some places than at others. Even though opinions about groups are subjective responses, the responses may be related to objective criteria. The average number of scientists in each group is 12.5 for high prestige, 10.4 for middle prestige, and 5.2 for low prestige. A rank order correlation between the total number of HEP affiliated

TABLE 1

RATED DEPARTMENTAL PRESTIGE

Rated Departmental Prestige	Departments Number	(Percent)	Affiliated HEP Number	(Percent)	Mean Number HEP for Each Department
High	4	(20)	50	(31)	12.5
Middle	5	(25)	52	(33)	10.4
Low	11	(55)	57	(36)	5.2
Total	20	(100)	159	(100)	8.0

This table applies only to HEP affiliated with universities. There are actually twenty-seven university departments in twenty universities; however, in those universities having HEP in two departments, the groups were rated as one department.
The range of mean prestige scores for each prestige category is: High, 3.54–4.43; Middle, 3.16–3.49; Low, 2.00–3.05.

with a group, including those few HEP not in the study, and its prestige rating was computed. The resulting correlation of .89 suggests that size probably is a major factor involved. It could be argued that prestige precedes size of research group, but that is asking this question: Why do some departments have HEP research while others do not? If the more prestigious departments in general were the first to obtain HEP research and subsequently increased the size of their research groups, then the explanation of prestige preceding size would be tenable. Detailed historical research, of course, would be necessary to support or reject that hypothesis.

Absolute size itself does not necessarily determine quality but it does increase the probability for diversification of research effort among all types of HEP research. Research diversification is related to prestige. In the high prestige group, both types of experimental research and also intermediate theoretical work are conducted in every department. In the middle prestige group, only counter/spark chamber work is done in all five departments with only three of those departments having bubble chamber research teams. (Two of the latter three also have two different types of theoretical research.) In the low prestige group, only one department out of eleven has both types of experimental research. Six departments have only one type of experimental research; however, five of the six also have theory research. Bubble chamber physicists are the largest group in the high prestige departments; counter/spark chamber physicists are the largest group in the middle prestige category; intermediate theorists are the largest group in the low prestige category.

If there is a prestige system, and the data suggest there is, what are the consequences for HEP working in such research environments? One correlate of prestige is the number of scientists who visit universities to give a seminar or talk about their research. Whether HEP from other institutions are more likely to visit high prestige departments or whether departments visited by a large number of HEP are more likely to be visible and thus highly rated is unclear. Although there probably are feedback effects, it seems logical that visitors to departments represent an effect of prestige. A large and prestigious research group is more likely to be able to afford to bring in guests, and guests are more likely to be willing to travel to prestigious groups.

Scientists at the high prestige departments report a much larger number of visiting scientists (79 percent reported that twenty or more visited during the previous year). HEP at the middle and low prestige departments, however, report that few visitors came and held seminars or gave talks. The reports showed more variability among the scientists in these institutions, which suggests uncertainty on their part. When asked about the number who came to his department, an experimentalist (at a middle prestige department) replied: "Not enough, I'm afraid. We're a bit isolated here at ———. Not many people want to come here, I feel."

Each scientist was asked the approximate amount of time he spent each week on research during the academic year. The average number of hours is greater in the high prestige departments (64 percent spent over thirty hours), somewhat less in the middle (61 percent spent over thirty hours), and considerably less in the low prestige departments (39 percent spent over thirty hours). The correlation between prestige scores and number of hours spent on research explained little of the variance (r=.29).

There is an inverse relationship between the proportion of the work week spent on teaching and the prestige of the department. At high prestige departments, 58 percent of the HEP spend less than one-third of their time teaching, whereas at middle prestige departments 43 percent spend that small amount, and at low prestige departments only 35 percent spend that little time on teaching.

Related to size and distribution of research effort is the fact that large sums of money are necessary to carry out experimental work. The distribution of funds gives an indication of the potential resources HEP have at their disposal. Counter groups use laboratory facilities, and funds are not directly given to the research teams; thus, regardless of where a counter experimentalist is affiliated, the research funds chargeable to him would simply be approximately his per capita share of the laboratory expenditure. This does not mean that groups spend the same amount of funds but that accounting for these funds is very difficult because most of the funds are disbursed internally at the laboratories.

Theoretical research, except for phenomenological computer analysis, essentially requires only travel funds and library facilities. Considering bubble chamber expenditures for the two years 1966–68, because these funds are granted directly to the groups (and assuming equal accelerator use), much greater sums (39,029 pounds) were spent per HEP at the high prestige departments, less (27,902) at middle prestige departments, and still less (13,232) at the low prestige departments.[6]

This suggests that HEP at high prestige departments

are able to command more research funds, and if funds are important in getting one's research done, then affiliation with a high prestige department provides important advantages. Some argument could be made that departments were originally awarded funds because of their prestige, and that prestige is therefore the cause rather than the result of funds. The Science Research Council has a policy, in fact, of supporting ten departmental bubble chamber groups (plus one more at the Rutherford Laboratory), and these have been considered by the council as major and minor centers. To some extent, then, the rating of departmental prestige by HEP may be partially a result of what is essentially an institutionalized governmental policy.

HEP were asked about funding in their departments to determine the perceptions of their relative situation vis à vis other departments. One HEP at a high prestige department rather surprisingly said:

I think we suffer a bit from being ——— University. We're one of the largest groups in the country, and it's difficult to persuade the paymaster that we ought to get the larger share. Many of them [officials] feel that ——— University can look after itself. [They] want to promote less effective departments to get going. I think this is a mistake myself.

In contrast, a HEP at a middle prestige department said:

I suspect we do slightly better than the average. There are places that do equally well; ——— University [the university in the above quote] does very well, very well, indeed. Do you know "X," the Prof there? Not surprising they do very well [laughter]. He's just exceptionally impressive. [He gets the money?] Yes, that's right. So is our boy [the department head] very impressive.

Interestingly enough, these remarks were somewhat prophetic of the outcome. Scientists at the middle prestige departments are more likely to say they are better off financially than most other groups: only 2 percent thought they were less well off while 26 percent in the low prestige group thought they were less well off and 19 percent in the high prestige group responded similarly. The concentration of various types of HEP that differ for various prestige groups might account for this finding, but type of HEP is not related to personal perception of relative financial position, so that could not account for the different perceptions.

Two factors probably explain this. In the middle prestige group, scientists probably have sufficient funds to carry out their research, and thus they feel they are well off. This factor has to do with expectations, illustrated by the first scientist quoted above who thought that his group should have more (and he is located in an institution that by any standard is well financed). A second factor involves a recent decision that has invoked questions of politics in the minds of some HEP at a high prestige department. One HEP said that in fact he thought HEP in Britain was well treated, but he added:

> I would complain about the way it's spent within high energy physics. I think the means by which finance within high energy physics is decided—[Science Research Council] committees and so on—I think we've been unjustly treated. In terms of financial support, ours is minute compared with some other groups. [On what basis is it determined?] I think there is one-to-one correspondence between the groups that receive heavy financial support and the members of the appropriate committees who decide how the money is spent—in fact, I know it's so. [You don't have a representative?] No. If and when we do, it will be different.

Another HEP suggested a similar reason why he thought his department was less well off than others.

> You've probably heard that our recent proposal to the SRC for ——— has been turned down. ——— University had a proposal for ——— [project] and large ——— [equipment], and they have a professor on the right committee. They could have used a smaller ———. There's a lot of bad feelings. One has noticed that the people on this board have managed to get

hundreds of thousands of pounds for their department. I'm sure there's a connection.

There is some evidence for this assertion. Departments with representation on the Science Research Council received approximately 90 percent of the bubble chamber funds for 1966–68, while those without representation received 10 percent. Whether representatives are chosen because departments already receive large amounts of funds or whether representatives help their departments receive large amounts of funds is a question that cannot be answered from my data.

If funds or prestige make any real difference in a scientist's career, it should be revealed in terms of his contributions to the scientific community and the recognition he receives from the community for performing the scientific role.

Scientific Productivity

The major contribution that scientists make to the community is research results presented as published reports. Hagstrom (1965) suggests that scientists *exchange* research findings for recognition from the community. Storer (1966) believes that, as in other social systems, the scientific community has something that the scientist wants, and the scientist has something that the community wants. Scientists want competent *response* to their creative efforts (research contributions), and the community wants the results of this research. In either way of looking at the reasons why scientists publish their work, the medium of exchange or the medium of creative expression is the published paper.

There are also other types of contributions. Teaching future scientists is looked upon as a worthwhile activity, although its value is difficult to measure. The actual number of Ph.D. theses directed might be a good indicator, if the number of theses did not depend more upon available staff than the actual individual contribution. And, of course, the amount of direction involved is a factor that would have to be taken into account.

Presenting papers at conferences is somewhat similar to publishing. One difference is that the work is communicated and therefore usable more quickly than in a published form. This contributes not only to quicker utilization but informs others of current interests and activities. Verbal presentation is a type of contribution that assists other scientists not only as to substantive content but also to strategy that they may wish to employ. Since most conference reports are later either published or included as part of a larger publication, they actually form a subset of publications rather than separate activities.

Each HEP was asked the number of papers, letters, books, and review papers (of research areas) that he had published or had had accepted for publication. Some of the scientists did not have such a list, and it was compiled during the interview from their reprint files or on occasion the scientist compiled his list and forwarded it later.

The cumulative total of publications that a HEP has produced is used as the measure of his scientific productivity. Letters and papers are not differentiated in the cumulative total because scientists tend to count them as equivalent. Although letters are shorter, they are a faster form of publication, and the urgency of the publication represents a value to the community that may offset the value of a longer, but slower, paper. Only eighteen HEP have published in book form, and these are counted as a publication equivalent to a paper.[7] These cumulative totals are used in correlational analysis and in regression analysis in which the residuals from the regression of productivity on professional age provide information on the scientists' "overproductivity" or "underproductivity," according to what he "ought" to have produced by that stage of his career. An index was devised dividing the group into high and low producers (see the appendix for details).

Scientific Recognition

Some activities that contribute to the community but are really indicative of recognition include refereeing papers, giving guest lectures and invited papers, serving on committees, and being editors or serving on the editorial boards of professional journals. These activities are services to the scientific community because they provide needed assistance in carrying out the overall program of the community; and yet it is questionable if any of these activities are voluntary. Scientists are asked to do them, and the basis for being asked is the assumption that the person is qualified to carry out the task. (Some political considerations may be involved, but that is another question.)

Giving lectures to other scientists and students confers a certain status on the lecturer.[8] Refereeing papers to decide which may or may not appear in journals; serving as an editor or on an editorial board, "gatekeeping" as it is called (see Crane, 1967); and serving on committees that decide questions or make policy regarding the future growth or direction of science all involve power (decision-making). Since the exercise of power is a source of pleasure to many people, receiving recognition by being asked to serve probably far outweighs the trouble that the service requires.[9] In addition to those activities that some might argue are both recognition and contributions, there are rewards that are clearly recognition: prizes given for outstanding accomplishments; membership in honorary societies; and being elected to office in professional societies.

A scoring procedure enabled recognition to be correlated with other measures, and the residuals from the regression of recognition scores on productivity provide information on whether the scientists have received as much recognition as they should for their research productivity. An index was also constructed dividing the scientists into groups with high and low recognition scores (see the appendix for details).

Correlates of Productivity

Several variables were analyzed for their effects on productivity. There is no significant relationship between productivity and (1) social class origins, (2) type of secondary school attended, (3) the type of undergraduate university attended, (4) class of undergraduate degree, or (5) type of university attended for the Ph.D.

At the Ph.D. level, several variables concerning the thesis adviser were examined: the status of the adviser,[10] the closeness with which the HEP worked with his adviser, and the contact maintained with the adviser since receiving the Ph.D. The possibility of influence other than the adviser's influence on the HEP while he was a research student was also considered. None of these variables were related to productivity: (1) the status of the adviser, (2) whether the HEP worked closely with his adviser, (3) had remained a personal friend of the adviser, or (4) was or was not influenced by other staff members.

Although the current prestige of departments may be used as an approximate measure of the prestige of the department from which the HEP took their Ph.D.s, the ratings are more indicative of the current ranking and are not necessarily the best indicator of an earlier ranking. It is the case, however, that only seven years have lapsed, on the average (median=six years), since the HEP took their Ph.D.s. The prestige rating of the Ph.D. department has no effect on scientific productivity.

The correlations between several independent variables and productivity are shown in table 2. Although depart-

TABLE 2

PRODUCT-MOMENT CORRELATIONS BETWEEN CERTAIN INDEPENDENT VARIABLES AND NUMBER OF PUBLICATIONS

Independent Variables	Number of Publications
*Prestige of Ph.D. university**	.07
*Prestige of present affiliation***	.24
*Proportion of time spent on research****	.13
*Number of hours weekly spent on research****	.27
*Proportion of time spent on teaching****	.04
*Chronological age****	.56
*Degree class of undergraduate degree*****	.11
*Professional age****	.65

*Based on 176 cases.
**Based on 159 cases.
***Based on 203 cases.
****Based on 169 cases.

mental prestige and amount of funds per bubble chamber scientist are related (realizing of course that only about one-fourth of the sample are bubble chamber specialists), and prestige of department and proportion of time available for research (with the complement of less time spent on teaching) are related, the correlations in table 2 do not indicate strong relationships between productivity and these variables. Being presently affiliated with high prestige departments, however, does not mean that any effects of these factors would necessarily have been great. To measure accurately the effects of these variables would require a sufficient period of time for any differences to have an effect. The one main determinant of productivity is professional age—essentially the number of years that a HEP has been conducting research. The prestige of current affiliation correlates with productivity .24.[11] That correlation can be examined more closely by looking at the percentage of HEP categorized as high or low productivity located in the various categories of departmental prestige, as well as examining the results of the analysis of variance (see table 3). It is the case that the mean number of publi-

TABLE 3

PRESTIGE OF CURRENT DEPARTMENT AFFILIATION: PRODUCTIVITY, MEAN PROFESSIONAL AGE, MEAN NUMBER OF PUBLICATIONS, AND MEAN RESIDUALS (FROM REGRESSION OF NUMBER OF PUBLICATIONS ON PROFESSIONAL AGE)

	Prestige of Current Departmental Affiliation			
Variables	Low	Middle	High	Total
Productivity				
Low	74	65	58	66
High	26	35	42	34
Total	100	100	100	100
(Base)	(50)	(52)	(57)	(159)
Mean professional age	6.91	8.48	7.84	7.01
Mean number of publications	12.65	16.42	18.52	14.71
Mean residuals (number of publications on professional age)	− 1.91	− .67	2.60	0.00
Statistics				

X^2 (*Productivity index*=2.94: 2 *df* , *prob.* <.30; *gamma*= .23)
F (*Professional age*= .99: 2,158 *df.*, *NS*)*
F (*Number of publications* =2.26: 2,158 *df.*, *NS*)
F (*Residuals* =2.22: 2,158 *df.*, *NS*)
$r_{productivity/prestige \cdot professional\ age} = .19$

*F ratios not significant at less than the .05 level are labeled "NS."

cations is greater for the high prestige categories, but it is not significantly greater. Although the percentages categorized "high" or "low" productivity vary in a positive direction with prestige, the productivity index is not greatly sensitive to extreme values. For example, in any given professional age group, a HEP with a very large number of publications might be classified "high" along with another whose number of publications was considerably fewer. These findings suggest that the prestige of a HEP's current affiliation has only a small—if any—effect on his productivity. Another variable will be considered, however, before concluding that the main effect on productivity is a scientist's professional age.

Scientists at Oxbridge and London universities are more productive than HEP elsewhere, while those HEP affiliated with the laboratories are slightly above the average (see table 4). This table partially explains the low correlation between percentage of time spent on research and productivity that was mentioned earlier. The HEP at the laboratories spend essentially all of their time on research, although part of their time is used to assist the university groups that do experiments at the labs and to serve on committees that help to administer various aspects of the labs' operation.

The extremely high number of publications by HEP at Oxbridge universities results in a lower correlation between publications and professional age, while the extremely low number of publications of the HEP at Scottish universities accounts for their lower-than-average correlation. At the same time there is a close correlation between professional age and publications by the HEP at London and Redbrick universities and the laboratories. These findings suggest that professional age and type of university are the two main variables in predicting number of publications. The question is: What is the nature of the various types of university departments that provide opportunities for different productivity? Rather than suggest an answer to that question at this time, I will present further analysis that will put the question in a different perspective. Since scientific productivity is only one aspect of the reward system, I will now consider recognition in terms of its relationship to the variables discussed for productivity.

Correlates of Recognition

If the universalistic norms of science are adhered to, any kind of social or educational background is irrelevant to the value of a scientist's work and the recognition he receives for it. The reward system in British HEP operates in a universalistic fashion. Social class origins, type of secondary school attended, or type of undergraduate university, are not related to recognition.

TABLE 4

TYPE OF INSTITUTION OF PRESENT AFFILIATION: PRODUCTIVITY, MEAN PROFESSIONAL AGE, MEAN NUMBER OF PUBLICATIONS, AND MEAN RESIDUALS (FROM REGRESSION OF NUMBER OF PUBLICATIONS ON PROFESSIONAL AGE)

Variables		Present Affiliation					
		Oxbridge	London	Redbrick	Scottish	Laboratory	Total
Productivity	*Low*	67	50	69	84	68	66
	High	33	50	31	16	32	34
	Totals	100	100	100	100	100	100
	(Base)	(30)	(36)	(74)	(19)	(44)	(203)
Mean professional age		9.33	8.31	6.35	9.37	4.45	7.01
Mean number of publications		22.17	19.83	12.01	12.26	11.05	14.71
Mean residuals (number of publications on professional age)		3.92	3.02	− 1.64	− 6.27	.47	0.00
Product-moment correlation: professional age and publications		.53	.74	.71	.46	.74	.65
Statistics							

$X^2 = 7.33$ (4 *df.*, *prob.* $< .20$)
F (*Professional age* $= 5.60$: 4,198 *df.*, *prob.* $< .001$)
F (*Number of publications* $= 5.35$: 4,198 *df.*, *prob.* $< .001$)
F (*Residuals* $= 4.09$: 4,198 *df.*, *prob.* $< .01$)

The one exception to these findings (see table 5) is that HEP with first-class undergraduate degrees are more likely to obtain recognition for their productivity than HEP with second-class degrees. The effects of degree class may result from a status consideration of undergraduate accomplishments. It will be recalled that many British academics do not take Ph.D.s, and in these cases the class of the undergraduate degree is significant in British academic circles. Furthermore, it may be that ability as measured by the first-class degree results in more significant research. Finally, it is possible that research done by a first-class degree scientist may appear to be of higher quality than work done by a second-class degree HEP. While all these hypotheses seem plausible, I will show that there is still a more plausible explanation that does not violate the norm of universalism.

The relationship between recognition and prestige of

TABLE 5

RECOGNITION AND PRODUCTIVITY BY CLASS OF FIRST DEGREE AND MEAN RESIDUALS (FROM THE REGRESSION OF RECOGNITION ON NUMBER OF PUBLICATIONS AND RECOGNITION SCORES)

	Class of First Degree				
	Second Productivity		First Productivity		
	Low	High	Low	High	Total
Recognition					
Low	83	58	72	41	66
High	17	42	28	59	34
Total	100	100	100	100	100
(Base)	(36)	(12)	(76)	(44)	(168)
Mean recognition scores	1.42		3.05		2.66
Residuals (recognition on number of publications)	−1.06		.18		0.00
Statistics					

X^2 (*Table*) = 18.69 (3 *df.*, *prob.* < .001)
X^2 (*Second class*) = 1.93 (1 *df.*, *prob.* < .20); *gamma* = .56
X^2 (*First class*) = 10.29 (1 *df.*, *prob.* < .01); *gamma* = .58
F (*Recognition scores*) = 10.08 (1,167 *df.*, *prob.* < .01)
F (*Residuals*) = 6.39 (1,167 *df.*, *prob.* < .05)

current affiliation is even less than the very weak relationship reported between prestige of current affiliation and productivity (table not shown). Productivity is related to *type* of institutional affiliation as shown earlier, but recognition is not related to institutional affiliation (see table 6). Recognition scores vary between institutions, but they are not significant; the residuals from the regression of recognition on number of publications are also not significantly different. This raises the question of the relationship between productivity and recognition, because if productivity is significantly different between institutions —and since recognition and productivity should be related—it would be expected that recognition would be different.

Relationship between Productivity and Recognition

There is a strong relationship between productivity and recognition. Table 7 shows this relationship for the productivity and recognition indices and includes Crane's (1965) data for comparison. It should be emphasized that the similarities in this table do not extend to her findings in general—especially since her analysis indicated that institutional prestige interfered with a universalistic distribution of rewards. In addition to the productivity and recognition indices, the product-moment correlation between professional age and number of publications (shown in table 2) is .65. Professional age thus explains 42 percent of the variance in number of publications. The correlation between professional age and recognition is .53, which means that professional age explains 28 percent of the variance in recognition. The correlation between publications and recognition is .69, which suggests that the number of publications is a more important factor in recognition than professional age. Since all three variables are related, a partial correlation of .53 between number of publications and recognition, controlling for profes-

TABLE 6

RECOGNITION BY INSTITUTION OF PRESENT AFFILIATION: RECOGNITION SCORES AND RESIDUALS (FROM THE REGRESSION OF RECOGNITION OF NUMBER OF PUBLICATIONS)

		Present Affiliation					
		Oxbridge	London	Redbrick	Scottish	Laboratory	Total
Recognition	*Low*	63	58	69	79	59	65
	High	37	42	31	21	41	35
	Total	100	100	100	100	100	(100)
	(Base)	(30)	(36)	(74)	(19)	(44)	(203)
Recognition scores		3.90	3.83	2.16	2.00	2.02	2.67
Residuals (recognition on number of publications)		− .33	.09	.06	− .16	.12	0.00
Statistics							

$X^2 = 3.54$ (4 *df.*, *prob.* $< .50$)
F (*Recognition scores*) $= 2.02$ (4,198 *df.*, *NS*)
F (*Residuals*) $= 0.13$ (4,198 *df.*, *NS*)

TABLE 7

PRODUCTIVITY AND RECOGNITION FOR TWO SAMPLES OF SCIENTISTS

British High Energy Physicists

	Productivity		
	Low	High	Total
Recognition			
Low	76	44	66
High	24	56	34
Total	100	100	100
(Base)	(135)	(68)	(203)

$X^2 = 18.29$ (1 *df.*, *prob.* $< .001$)
gamma = .59

Crane's Sample of American Scientists*

	Productivity		
	Low	High	Total
Recognition			
Low	70	44	56
High	30	56	44
Total	100	100	100
(Base)	(69)	(81)	(150)

$X^2 = 8.83$ (1 *df.*, *prob.* $< .01$)
gamma = .48

*Source: Crane (1965) n. 18, p. 702; and n. 47, p. 710.

sional age, indicates that, regardless of age, a HEP receives recognition for his contributions.

This finding is somewhat contrary to a notion expressed frequently during preliminary interviews with British scientists in the United States. The prevalent notion was that age is an extremely important factor in Britain in obtaining recognition—that is, age is not a *sufficient* but a *necessary* condition in being able to assume scientific responsibilities and presumably, therefore, win recognition.[12] If age is not a strong influence, other variables must be considered in order to answer these unexplained questions:

Since productivity and recognition are related, why do

HEP with first-class undergraduate degrees obtain more recognition when they are not significantly more productive?

Since productivity and recognition are related, why are HEP significantly more productive at some institutions while their recognition scores are not significantly different, but at the same time the mean residuals from the regression of recognition on number of publications are similar at the various institutions? The answers to these questions are not found in the institutional or social environment outside of science, but result from the division of labor in the HEP community.

The Nature of Science and the Reward System

Although scientists are rewarded for their scientific contributions regardless of social factors in the whole society, certain status differences exist. These differences result from a division of labor in science. Hagstrom (1965:245) states that "Functionally differentiated elements are rare in science, and the only really well-developed example is the differentiation of physics into 'theoretical' and 'experimental' components." Hagstrom sees two conditions for such differentiation. One is the logical elaboration of abstract theory that requires special training in order to deal creatively with theories. The other condition for this differentiation is the growth of technological complexity in the collection of data by experiments and field research. Again, special training is necessary to deal with these complexities.

Hagstrom (1965:247) mentions, furthermore, that the role of the theorist in contemporary physics is almost completely differentiated from the role of experimenter, and "the differentiation is most complete in nuclear physics." Only one HEP indicated he had done some theoretical work, and no theorist indicated having done any experimental work. It could be expected, therefore, that this role differentiation would result in some measurable status differences.

The first question regarding the differences in recognition received by the first-class undergraduate degree holders is explained by the fact that theorists are more likely to obtain first-class degrees. Of the 56 theorists with classed degrees, 89 percent got a first-class degree; of the 112 experimentalists, only 62 percent obtained a first (chi-square=11.85, 1 df., prob. < 001). Theorists are significantly younger than experimentalists, but productivity and recognition scores are not significantly different between the two types of scientists, which means that theorists produce slightly more papers for their age. The expected productivity based on professional age does not differ significantly, but the amount of recognition received for the number of publications is significantly more for theorists (see Table 8). In short, theorists obtain more recognition

TABLE 8

TYPE OF HEP AND SELECTED VARIABLES

Variables	Type of HEP: Experimentalist	Type of HEP: Theorist	F Ratio	Significance Level*
Professional age	7.62	5.89	4.44	.05
Mean number of publications	14.95	14.26	0.11	NS
Mean recognition scores	2.36	3.31	2.55	NS
Mean residuals (number of publications on professional age)	− .68	1.37	1.73	NS
Mean residuals (recognition scores on number of publications)	− .39	.73	6.59	.05
Product-moment correlations: recognition scores and number of publications	.61	.72	...	...
Partial correlations: recognition scores and number of publications controlling for professional age	.34	.61	...	...

*NS indicates the F ratio is not statistically significant at the .05 level (lower levels of significance are generally not provided in reference tables) but does not eliminate substantive significance.

than experimentalists for equivalent publication. For each type of scientist, productivity and recognition are still related, but the relationship is stronger for theorists. This is shown by the partial correlations between number of publications and recognition, controlling for professional age, which are .61 and .34 respectively for theorists and experimentalists. Of course, one possible explanation for this finding may be an artifact of the recognition measure used. Three of the items concern lectures and invited papers. Experimentalists work in groups, and it might be that only one of the group would need to lecture on that group's research. Since a theorist tends to work more often by himself or with only a few collaborators, reporting his research results means that he personally is more likely to give the lecture. Each scientist would be equally eligible for the other items if in fact the scientific community did not put greater value on theoretical contributions. The question is, why do they receive more recognition?

One possible explanation is that theorists have a broader perspective. They have an interest in more general problems in physics because ideas for theories may come from a variety of sources, and the theories constructed should provide explanations to cover a broad area. Experimentalists, in contrast, are interested in specific types of experiments rather than being concerned about all of elementary particle physics. This is a result, in part, of the approximate two-year time span to complete experiments, a relatively new time factor that results from the complex technology involved in Big Science. Further elaboration of the effects of the role differences will be presented in discussing the second question.

The second question that remains unanswered is: Since productivity and recognition are related, why are HEP significantly more productive at some institutions but not significantly more recognized, while at the same time the mean residuals from the regression of recognition on number of publications are similar at the various institutions? To answer this question the type of HEP should be controlled when looking at present affiliation—not because theorists are more productive than experimentalists when viewed as two groups, but because the data show that the division of labor in the scientific community makes a difference in recognition and that difference should be further examined. Table 9 provides the data to examine this question.

Divided into institutions and type of HEP, there are differences in the productivity residuals between types of HEP and institutions, and there is also an interaction effect. Theorists are, in general, more productive when holding affiliation constant; scientists at London University and the laboratories are more productive when holding type of HEP constant. The interaction effect is present because Oxbridge theorists are significantly more productive than Oxbridge experimentalists, while the London experimentalists are more productive than the London theorists. The explanation for these differences is not obvious. For example, time to do research is not the main reason for the labs' "overproductivity" since, in general, time for research and productivity are not related—because every HEP has sufficient time even though some have more time than others. The best explanation involves the historical emphasis in certain institutions for certain types of theoretical or experimental research. These institutions have been able to attract and retain some of the more highly productive individuals. The general atmosphere of the institution possibly could explain some of the differences, but in the case of the Oxbridge universities the general climate for research would not explain the differences between the experimentalists and theorists. The explanation seems to remain that these universities have traditionally been strong in mathematics and theoretical physics but have not had equivalent facilities for experimental research in elementary particles. An explanation in more detail would require an

TABLE 9

TWO-WAY ANALYSIS OF VARIANCE OF REGRESSION RESIDUALS FOR TYPE OF HEP BY INSTITUTION OF PRESENT AFFILIATION

		Institution of Present Affiliation				
Residuals	Type of HEP	Oxbridge	London	Redbrick	Scottish	Laboratories
1. *Regression of number of publications on professional age*	*Exp.*	− 2.35	3.21	− .99	−7.76	.08
	Theorists	18.55*	2.53	−2.48	−4.22	1.51
2. *Regression of recognition scores on number of publications*	*Exp.*	− .74	− .15	− .51	− .94	− .03
	Theorists	.64	.74	.80	.92	.51

F ratios

1. *Type of HEP* = 8.71 (1,193 *df.*, *prob.* <.01)
   Institution = 8.54 (4,193 *df.*, *prob.* <.001)
   Interaction = 6.62 (4,193 *df.*, *prob.* <.001)
2. *Type of HEP* = 5.65 (1,193 *df.*, *prob.* <.05)
   Institution = 0.07 (4,193 *df.*, *NS*)
   Interaction = 0.22 (4,193 *df* , *NS*)

Note: The number of HEP in each of the above cells is different but the program used corrects for unequal cell size. The actual cell sizes are as follows:

21	26	42	11	32	Total = 132
9	10	32	8	12	Total = 71

*This figure results from a small group of highly productive theorists.

historical account involving names and specific institutions that would violate my promise of anonymity.

In general, the Scottish universities have not been as strong in research as the Oxbridge and London universities, while the Redbrick universities have been more successful in research than the Scottish universities but less productive than Oxbridge and London.

The lower part of table 9 indicates that, regardless of affiliation, theorists receive more recognition, while at the same time experimentalists receive less. It is particularly interesting that theorists at Scottish universities have, on the average, slightly more recognition than theorists elsewhere, while their productivity expected from professional age is the least. After theorists at Scottish universities, in decreasing amounts, theorists at Redbrick, then London, and then Oxbridge have "excess" recognition. Although those differences are not statistically significant, they give clues for another type of investigation. Since the distribution of types of research varies by institution, and since there is a slight difference in recognition residuals, the residuals for the various types of research specialists should be examined (see table 10). While the productivity residuals are clearly not significant, they show a trend that explains in part the differential productivity of HEP at the various institutions. Counter/spark chamber physicists are the only group to have a negative mean residual. The increase then goes from bubble chamber physicists to phenomenologists to intermediate theorists, and finally to abstract theorists. Abstract theorists have the highest mean residual, and it is relevant to indicate that not only is it possible for abstract theorists to prepare a paper in the least amount of time but that Oxbridge universities have several abstract theorists on their staffs—which accounts, in part, for the very high level of productivity for theorists at the Oxbridge universities.

The recognition residuals are more interesting than the productivity residuals. Although they too are not signifi-

TABLE 10

ANALYSIS OF VARIANCE OF REGRESSION RESIDUALS FOR TYPE OF HEF

Residuals	Type of HEP: Experimentalist		Theorist		
	Counter Spark/chamber	Bubble Chamber	Phenomenologist	Intermediate	Abstract
1. *Regression of number of publications on professional age*	−1.32	.29	.55	.64	.66
2. *Regression of recognition scores on number of publications*	− .35	−.46	1.19	.85	−.74
F ratios					

1. $F = 1.23$ (4,198 *df.*, *NS*)
2. $F = 2.32$ (4,198 *df.*, *NS*)

cant, the *F-ratio* would only have to be .13 more to be significant at the .05 level. A lack of statistical significance, however, does not remove the indications of substantive importance. The two groups with positive mean residuals are phenomenologists and intermediate theorists—abstract theorists are the most underrecognized group in the whole community. These results extend the explanation above of why theorists are more likely to receive recognition. If theoretical contributions are more valuable to the HEP community, then contributions of intermediate theorists are clearly important and contributions of phenomenologists are even more important. Phenomenologists deal with the actual experimental data in constructing phenomenological models of the "behavior" of the particles. Since they are writing about the experimentalist's real world, experimentalists are able to benefit from these theoretical contributions whereas intermediate theoretical results are of little direct benefit to them and it is doubtful that the abstract theorist's work is of any benefit to them.

It is possible that the phenomenologists' status in elementary particle physics at the present time results from the lack of adequate abstract theories. The suggestion that phenomenological theories are always the most beneficial or receive the most attention is not being made—rather, what is being suggested is a functional theory of valuation. Those theories that have the most potential benefit for any given science at a particular time are the ones that will be the most acclaimed.

The Cole and Cole (1968) research on American physicists suggested that "visibility" in the scientific community is determined by four strong determinants: quality of research as measured by citations to research; honorific rewards received; prestige of university department; and specialty.[13] Physicists whose research is elementary particles were the most visible in their study.

While Cole and Cole used as a measure of visibility the proportion of physicists from a sample who are familiar to some extent with a scientist's work, that measure is not available for the present study. For the last several years, however, the center of high energy physics research has been in the United States and more recently the European center has been at Geneva.

The question then was whether HEP who have spent some time in a position out of Great Britain—and therefore have potentially enhanced their visibility among their colleagues—increased their chances for recognition (or productivity)? Comparisons of HEP who have held a position only in Britain, in the United States, elsewhere out of Britain, or both of the latter, indicate that this is not the case. There may be other considerations that are involved in the awarding of recognition by the scientific community, but none of the variables analyzed indicate that particularistic criteria are important in the allocation of recognition for the contributions made to the HEP community.

Of great interest is the difference found between the reward system in the United States and Britain. It is, of course, possible—but not likely because of the similar organization of all research effort—that the HEP community is exceptional. One of the reasons why previous research has not found some of the differences between experimentalists and theorists is because this division of labor is not so obvious in many disciplines—but even where it is obvious, no results based on these differences yet have been emphasized.

As the division of labor in science occurs as a result of increasingly sophisticated methodology and complex theory, the future developments of science are likely to result in the same status differentiation with its potential problems of the equitable allocation of rewards. The problems are potential since retention and recruitment for the scientific community would suffer if it appeared that only a few scientists could obtain the recognition that must come after applying one's self to the difficulties and frustrations of research. This raises the question of whether problems of originality and competition are different for the scientists engaged in the various types of theoretical or experimental research.

Competition in Science

5

Science as a social activity involves several social processes. *Cooperation* in science is represented by collaboration and mutual criticism.[1] *Cooptation* in science has not been systematically studied, but a novel by William Cooper about British scientists entitled *The Struggles of Albert Woods* (1953:32)[2] has a good example of this process. "This is not the moment for me to go into Didbin's motives," the narrator says; "all I will observe in passing is that if you feel a young man is likely to be a rival in your own line [of research] it is not a bad idea to send for him to work with you." *Discrimination* is against the norms of science, and although it rarely occurs overtly, one example is the discrimination Jewish scientists suffered prior to World War II.[3] *Competition* in science has been investigated on two levels: institutional competition and competition experienced by the individual scientists.

Ben-David (1965a) analyzed institutional competition when he asked: Why did the medical sciences flourish in Germany and the United States during the nineteenth and twentieth centuries in contrast to the lag in Great Britain and France? He found that scientists who became successful were rewarded by being given university chairs and facilities for research. Their success encouraged others to take up scientific research and help to establish the role of professional researcher. A major part of this process in the United States and Germany was due to the decentralization of the university systems in which competition was strong. Thus, for the growth and progress of a discipline interinstitutional competition may be a positive factor.[4]

On an individual level, competition causes scientists to work harder in an attempt to achieve or obtain more of the scarce item—recognition—from the scientific community. This fact of scientific life is vividly illustrated by Rutherford when he wrote his mother in 1902 that "I have to keep going, as there are always people on my track. I have to publish my present work as rapidly as possible in order to keep in the race . . ." (quoted in Crowther, 1952:54). Written when Henri Becquerel and the Curies were also doing important work in physics on the problems of radioactivity, this shows that scientists, with already well-established reputations, nevertheless still want the pleasure of reporting first the results of their scientific investigations. This does not distract from or criticize the motivations for scientific research, but rather it helps to confirm the long-held view that scientists want recognition for their research efforts. Darwin is quoted as saying that "My love of natural science . . . has been much aided by the ambition to be esteemed by my fellow naturalists" (quoted in Merton, 1962:455). Hans Selye has stated, ". . . all the scientists I know sufficiently well to judge (and I include myself in this group) are extremely anxious to have their work recognized and approved by others. Is it not below the dignity of an objective scientific mind to permit . . . a distortion of his true motives? Besides, what is there to be ashamed of?" (quoted in Merton, 1965:122). More recently, a contemporary scientist, apparently somewhat disgusted because a writer had called attention to this motivation, said when reviewing a book that mentioned the desire for recognition among scientists, "Yes, Virginia, scientists do love recognition, but only since Pythagoras" (Lederman, 1969:169).

In fiction this aspect of science is dramatized in Mitchell Wilson's *Live With Lightning* (1949:103–4), in which the (wholly?) fictitious Columbia University physicist, Erik Gorin, says to his wife Savina:

> ". . . I *am* ambitious, Savina. I've got it so bad I'm afraid if I once let it get started, it'll run away with me. A long time ago, I told you that I ache with wanting success. . . . It's not that I want to be famous or rich. What I want, what I burn for, what I'd give damn near anything in my life for, is to be good enough to deserve being famous, as a scientist.
>
> ". . . Love is fine, darling, but I tell you it's not like that, and that's what I want. Even on the smallest scale. God, God, God, how I want it!"

If unlimited recognition were available to all seekers, competition would obviously not exist. One positive consequence of competition is that when several individuals put forth their best efforts toward achieving a scarce item that most of them cannot win, the result is a higher average achievement level for all participants in the relevant activity. Hornig (1969:527) writes that in science, ". . . as in economic processes, competition stimulates the quality of performance and must be fostered, together with the cooperation which comes through an open, widespread, and effective communication system among scientists."

In research, competition may keep enthusiasm alive to enable the unattractive aspects of the work to get done. A group leader said that "while physics may look like a very attractive subject, in reality it's 90–95 percent drudgery." Competition helps to motivate scientists who otherwise might put less than their best efforts into it. One experimentalist said:

> I have suspicions that less than one physicist in three, that I've met, is really and deeply and drastically motivated by a desire

to "find out." In one physicist in three, you'll find he's the kind who will lie in bed at night wondering about whether a certain cross-section is 34 or 36 millibars, but the remaining 60 to 70 percent will not. Large numbers of us, as myself, very much enjoy doing experiments and analyzing them. I'm never that particular about what experiment I'm doing as long as it presents me with some interesting and challenging problems to meet on the way. I think those that reach the top in this field are those that are strictly physics-motivated. These are the people who will not let themselves get too deeply involved in such things as [computer] programming. I'm quite sure it's concealed, but I think there's a fair level of dissatisfaction probably. Research is not what it's made to be.

This kind of sentiment coupled with the fact that much of the work is drudgery ("donkey work" as HEP call it) substantiates the idea that competition has some positive effects. A young experimentalist suggested that competition was good because it

. . . keeps the physicists a bit more interested in it. At times, it can get humdrum. In the middle of an experiment if you can engender a little excitement other than physics, I think this is quite good. This trying to publish first—which is decried a bit—keeps people on their toes somewhat.

The negative consequences of competition may result when failure to reach a goal either discourages an individual from further competition or causes him to adopt a different mode of operation to get the edge or advantage over his competitors. "Deviant" behavior results when a circuitous approach to a goal is used rather than following the "rules of the game."

While Ben-David's research has focused on the positive aspects of systemic competition, Merton and Hagstrom's research has focused on competition at the individual level with special attention to the consequences of individual competition for recognition from the scientific community. These consequences of competition take the form of such deviant behavior as hasty publications, fraud or theft, and secrecy.

Merton's Research on Competition

Merton's (1957; 1961; 1963; 1965) interest in the competitive aspects of science has been to look historically at multiple discoveries and the priority disputes that arose as a result. These disputes show that scientists are interested in recognition, and that when it is not forthcoming for making a scientific contribution, or when the recognition due is disputed by an earlier claimant to the discovery, considerable energy is expended on marshaling scientific friends to help settle the dispute. In such situations, the scientist faces a dilemma. The institution of science demands originality. Only original contributions are needed by the community, and it is for such original contributions that the community accords recognition. At the same time, scientists are sometimes hesitant to admit that they want recognition because the norm of disinterestedness includes a "humility" factor. Scientists should not proclaim that their research is great and earth-shattering. A good example of humility in science is shown in the words of Joseph Henry, inscribed in the foyer of John Green Hall at Princeton University:

> My life has been principally devoted to science, and my investigations in the different branches of physics have given me some reputation in the line of original discovery: I have sought no patent for inventions, and have solicited no remuneration for my labors, expecting only in return to enjoy the consciousness of having added to the sum of human knowledge.

The hesitancy among scientists and their biographers to admit that recognition is being sought caused Merton (1965:116) to propose a rule regarding biographies and autobiographies of scientists, stating that ". . . whenever the biography or autobiography of a scientist announces that he had little or no concern with priority, there is reasonably good chance that, not many pages later in the book, we shall find him deeply embroiled in one or another battle over priority." Most of Merton's paper is devoted to instances in support of this hypothesis with one of the most interesting cases being Ernest Jones's biography of Freud.

In their attempts to gain recognition, scientists are caught in a type of competition much different from the type of competition among other professionals. Businessmen can conceivably expand their markets or find new ones.[5] Literary men and artists are in competition with one another for readers and appreciation, but they do not compete on the performance of very similar or identical actions. Competition in science is like a race between runners on the same track and over the same distance at the same time. But even in track and field events, there are usually also "winners" with second- and third-place awards. This contrasts to scientific research where, although recognition is usually accorded to scientists making truly independent discoveries, the researcher who presents what is already known to the community may lose all or most of the recognition that would normally accrue to him. This results because scientists are trying to discover the truth about nature. As Price (1963:69) puts it, if Michelangelo or Beethoven had not existed, their artistic contributions would have been lost to humanity, but "If Copernicus or Fermi had never existed, essentially the same contributions would have had to come from other people. There is, in fact, only one world to discover, and as each morsel of perception is achieved, the discoverer must be honored or forgotten." Truth, importance, and originality are the marks of a valuable contribution to science. Only original research is regarded as a contribution, and a publication that does not have priority over other publications concerning the same problem is of secondary value.

Merton's work, based on historical data, provides ample evidence that scientists want recognition for their contributions.[6] Hagstrom's research, based on competition in modern science, discusses the prevalence and severity of competition and the consequences resulting from competitive situations when several people are trying to make the same discovery.

Hagstrom's Research on Competition

Hagstrom (1965) defines competition as the situation that exists when scientific research is likely to lead to the same discovery. When scientists are likely to be anticipated, competition is prevalent. Competition is more prevalent when scientists agree on the relative importance of the scientific problems and when there are many scientists at work on these problems.

Physics is a competitive discipline. It has a well-developed way of looking at physical problems. There is theory (or models) that can be tested, and there are very precise methods of collecting data. In comparison to physics, Hagstrom suggests, chemistry and molecular biology have less well-developed theories and experimental results are open to various interpretations. Finally, he suggests that while formal scientists (in mathematics, statistics, logic) have well-developed theories, there is greater freedom since an infinite number of mathematical systems is possible. Hagstrom (1965:74–75) hypothesized that the prevalence of competition would be greatest in physics and least in the formal sciences, with molecular biology coming in between.[7] His data on the percentages of each group that had experienced anticipation upheld this hypothesis.

Prevalence of competition is only one side of the coin. The severity of competition is also important. In addition to agreement on significant problems and a large number of people to work on them, Hagstrom suggests that the degree of precision that can be obtained in results is of extreme importance in determining the severity of competition. Consequently, Hagstrom (1965:75) hypothesized that severity of competition would be *greatest* in the formal sciences (because replication of proofs can be immediate when formal scientists are set upon the right track), *somewhat less* in the physical sciences, and *least* in the biological sciences. Severity is expected to be less in the biological sciences, as compared to the physical sciences, because of the lack of precision of methods and in the development of theories. Using the percentage of scientists expressing concern about being anticipated, Hagstrom's data showed that physical scientists and biological scientists were in the opposite positions predicted although the hypothesis held for the formal scientists.[8] These observations and tentative tests of hypotheses were based on small samples, but more recently Hagstrom (1966; 1967) extended his study of competition to a large sample of American scientists, the first study to describe the extent of competition and its correlates for a representative sample of American scientists. He considered the same hypotheses for this larger sample of 1,947 scientists.

The prevalence of competition is different for scientific fields in the large sample, but the differences between fields are not great. From the most competitive to the least competitive, using the percentages of each group having experienced anticipation as the measure of the extent of competition, he found the ranking from most competitive to least competitive as follows: chemistry, 68 percent; experimental biology, 67 percent; theoretical physics, 64 percent; experimental physics, 60 percent; other biology (which includes botany, zoology, anatomy, and ecology), 55 percent; and, mathematics and statistics, 54 percent. The main differences between the early and recent samples is that experimental biology and chemistry are more competitive than physics. Theoretical physics, however, is almost as competitive as chemistry and experimental biology.

Hagstrom used two measures of the severity of competition—concern about being anticipated and whether or not the scientist published as a result of being anticipated. Using the latter, the data showed that competition is most severe in physical sciences and least severe in the biological sciences (Hagstrom, 1967:8).[9]

Competition in High Energy Physics

According to the criteria that result in competition as outlined by Hagstrom—agreement on significant problems and a large number of people to work on them—high energy physics is a competitive specialty.

Agreement on significant problems requires a research paradigm described by Kuhn (1962) as "normal" science. Absence of adequate theory does not mean there is disagreement on the important problems. Agreement may exist that certain kinds of answers would be fruitful in *constructing* or *testing* a theory. In elementary particle physics, the consensus is that no adequate theory exists. There are models that have varying utility, but there is no theory that is really adequate.[10] An eminent theorist said:

> . . . We have tried to apply to high energy physics the general principles, if you like, from quantum theory, I mean quantum theory of the atom, etc., plus certain very plausible assumptions about the behavior of basic interactions and so on that seem very natural but are hard to prove. We haven't yet out of this got a complete set of fundamental laws that are sufficient to describe what's going on or to predict phenomena. . . . We don't have any contradictions. We have many things we can't explain, can't predict—but we can't even predict anything wrong! We just don't know. That makes it harder to guess where the answer might come.

Because an adequate theory does not exist is itself sufficient to cause competition among theorists in their attempts to try to create a new approach.

Hagstrom (1965:77) suggests that competition is more prevalent in theoretical than in experimental physics because theoretical physics usually requires no equipment of any kind and replications are unnecessary. Phenomenological research sometimes requires computers because large amounts of data are used to check and construct models, but that is not like experimental equipment that at times must be designed and built to order.

Experimental research has some characteristics that seem at the same time to make it both more and less competitive than theoretical research. It seems to be less competitive because replications not only may be more acceptable but also may be desirable. Experimental work seems to be more competitive than theoretical research because there are fewer problems on which to work. That is, if in the formal sciences there are an infinite number of systems that may be developed, in theoretical physics there are also a large—if not infinite—number of different ways to form a theoretical approach to explain phenomena. There are fewer problems for experimental research because of the limited number of variables involved. The group leader of a large bubble chamber research group in the United States was asked in preliminary interviews: "How frequently are similar experiments proposed to a machine committee?" To which he replied:

> Among non-bubble chamber groups [i.e., counter and spark chamber], about 5 to 10 percent of the time. Among bubble chamber groups, about 10 to 25 percent of the time. [Why do you suppose similar experiments are proposed so frequently?] It happens because there are a limited number of variables. The energy level is chosen, the target is chosen, and the beam. You get many experiments in bubble chambers at once [on the same film because it records all interactions], and the unintended findings are usually the most interesting. This can help to create possible competition since different groups may be looking for different things on the same film, and then each group might wonder if the other group is looking at or found what they see as something very interesting.

Another bubble chamber physicist has discussed the problem of competition in detail. Swatez (1966) studied the operation of the Lawrence Radiation Laboratory at

Berkeley, and Professor Louis Alvarez (the 1968 Nobel Prize winner for physics) responded to a sentence in the first draft of the research report that stated that "Some conflict among physicists is engendered by competition for the use of scanners" (149). Alvarez added that there were problems of possessiveness and competition in science. Formerly, physicists owned their own apparatus, data, and results. Then when Big Science came along, engineering notes for bubble chambers and scanning machines were made available to all, including commercial firms, which then built machines for sale. Everything was published. Computer programs were shared as well, ". . . if we hadn't, the smaller schools would have been severely handicapped, and we would have had most of the field to ourselves." Finally, he added,

> We sent film to many labs and, as I said, were scooped by them on several important discoveries. The Ξ^* and the ϕ were discovered in the K^- film we sent to UCLA, and the η (eta) were found by Johns Hopkins in the same way. In the end, it turned out that all a physicist can call his own is his own data on his own events; his own IBM printouts! And we often give our data to others, to include in "world summaries." Our people get pretty exercised if someone with twenty events asks our group to send its sixty events to be added, to give a "world total" of eighty events—the final paper to be published under the names of the people with twenty events—whose paper may include a credit line "We thank the Berkeley group for allowing us to include their data before publication." If one is asked, he feels obliged to give up all his hard-won data for a completely impersonal credit line. It is no wonder that feelings sometimes run a bit high. A human being has difficulty working long hours in a competitive atmosphere, when he feels constrained (on pain of being called a "bad guy") if he doesn't give all his hard-won data away. His professional future depends on his ability to publish "original papers" on "original data" [Swatez, 1966:149].

Hagstrom (1965:78) contends that competition in experimental research ". . . would always be most intense in research on such problems were it not for the fact that some workers have nearly a monopoly on the research facilities necessary for such work." Hagstrom believes oligopoly would be a better word since there is more than one accelerator at the highest energy ranges.

If a monopoly on machines was possible at an earlier time, it is not the case now. The CERN laboratory at Geneva has an energy potential (28 BeV) almost exactly that of the largest United States machine at Brookhaven (30 BeV). The Soviet Union has an even larger machine (76 BeV). The Batavia (U.S.) machine's impact is uncertain.

British scientists may work at CERN or on either of the two national machines, the Nimrod accelerator at the Rutherford High Energy Laboratory (7 BeV) and the Nina accelerator (4 BeV) at Daresbury Nuclear Physics Laboratory. But as with all machines at less than the highest available energies, HEP may be somewhat at a disadvantage. Although British scientists have CERN privileges (Britain contributes almost one-fourth of CERN's operating budget), most HEP work on national machines for reasons of convenience. There are overlapping energy ranges between accelerators, of course, and these intersections provide competitive possibilities. At higher energy accelerators, scientists have an advantage because of greater beam intensity, which refers to the number of particles per pulse of the accelerator.[11]

While 1–7 BeV is the same on any machine, the 1–7 BeV range of the higher energy machines produces more particles each minute. Thus to obtain one hundred events at a certain energy level, a higher beam intensity makes it possible to do the experiment more quickly. By simple analogy, the more bullets fired at a target, the higher the probability of hitting it. Should a topical set of problems be proposed, it is then reasonable to see how experimentalists with access to the higher energy accelerators will "skim off the cream" from the problem area.

It is the case that not all machines are in the same competition. Some accelerators initially produce protons while others produce electrons. Proton accelerators include, for example, the Berkeley, Brookhaven, Nimrod (at Rutherford Laboratory), and CERN machines. Nina (at Daresbury Nuclear Physics Laboratory), DESY (in West Germany), Cambridge Electronic Accelerator (joint Harvard-M.I.T.), and Cornell are locations of electron accelerators. Different research problems call for different machines, and scientists who work on a machine rather than a problem have to plan their research accordingly.

The type and energy of accelerators are not the only factors affecting competition. The style with which scientists go about their research in both a given social and economic matrix affects the speed with which scientists get out their research findings. An American physicist commented on national research styles in answer to a question about high energy physics research in Britain.

Support for research is not good. The government is too conservative. The style is limited by inhibitions. You see, there are different approaches to science. Italians are carefree, but the English are methodical.

This type of statement plus opinions of British scientists at American universities and laboratories indicated that HEP in Britain operate under several types of handicaps that might result in their being less able to compete as effectively on an international basis. Several suggested that financial support is inadequate.[12]

In general, it is poor, that is why many of us are over here. Its greatest defect is the lack of computers; as you probably realize, you need large computers to do this job because you do complicated computations and the one thing that stops you doing the work readily in England is the lack of computers. When I went back last time, we used to have to fly to Germany to get to the computers. The British computers were available, but we could only get them for commercial rates, which was completely out of the question. We didn't have enough money for that. The Germans had a computer for university use that they sold to us at cost rate. So we had to go there to do it. It has improved slightly but not very much in the last few years. [Is the Atlas computer at Rutherford workable for problems in high energy?]

It is, but the Atlas computer took a very long time to come on here. It didn't have a good compiler when it started, it broke down frequently, and didn't run the jobs very well. It was, by present-day standards, a bad computer. It was very expensive, and didn't live up to expectations.

Another British experimentalist said that his American group had never been beaten to publication by their rival Europeans. Asked how he managed that, he simply replied, "We're more competent." Competence seemed to be inherent in another scientist's thoughts about why he does better work in the United States. He remarked,

Here, there is a larger concentration of more good research people. My work is different. In ——— University, I was working on a small part of a small experiment that not very many people were interested in. Here, I'm working on a big facility that a lot of interesting people visit, and I come in contact with a large fraction of these.

Another British experimentalist in America commented on the general problem of both British national accelerators. Besides being late on the scene, the accelerators are both of relatively low energy. He said:

The northern part of England complained until they got Daresbury.[13] Now they have to almost use only that facility, even though it is essentially able only to do "gap" work.

He means that the cream has been skimmed, and filling in useful but somewhat unexciting data is the major function of the Nina accelerator.

Collaboration within Britain seemed to prevent internal competition and at the same time hinder British efforts at international competition. Regarding competition within Britain, a British experimentalist in America at the time said:

They're quite reasonable in England. There isn't a tendency like there is here to be working on the same field and try to publish before. If one has common interests, normally one wants to collaborate.

Collaborations may hinder international competition because they are unwieldy in terms of organization. Coordinating different groups to complete various aspects of an experiment (or analysis of a large number of filmed events) in a reasonable time and by the same precise procedures is difficult. While several smaller groups are cooperating on one experiment, a subgroup in the United States can be coordinated from a central location with more auxiliary machinery.

To summarize, British scientists indicated that financial resources were scarce in Britain, that research is handicapped by relatively low energy machines that have come on the scene late, and by having smaller groups with a scarcity of auxiliary facilities that necessitated collaborative efforts. These factors suggested the possibility that competition would be less in Britain and that consequently the scientists would be more likely either to work on noncompetitive fringe areas or experience anticipation to a considerable extent (especially at the hands of physicists in the United States). The second expectation was based in part on the various references that scientists made to one particular race for priority between a European collaboration (which included several British scientists) and an American group. That race was for the discovery of the Omega-minus particle.

Another case involved an instance of disputed priority. Both cases involve competition for recognition, and in both instances the advantage seemed to be with the American groups for the reasons mentioned above and subsumed under the general heading of superior research arrangements—not superior scientists!

Two Instances of International Competition: The Case of the Omega-minus and a Case of Disputed Priority

The search for the Omega-minus came soon after it was predicted by a well-known theorist. This race reveals the differences between the research atmospheres in Europe (including Britain, of course) and the United States, and suggests that research atmospheres may be more important than nationality of scientists in explaining research styles. A scientist working in a specific atmosphere may be caught up in it rather than being himself the prime cause of the atmosphere. One account of the episode was related by a British HEP who worked with the American group in the race to find the Omega-minus particle.

This case is particularly knocking to us. It was a direct test of a theorist's prediction. He said there should be a particle, and said what the mass was. And we looked like mad for it. We treaded blood in direct competition to CERN, because they had their beam[14] running before ours for several months. But they didn't have a bubble chamber running, or rather it was one of the nationalistic jokes. They had a chamber that was working, but that wasn't the chamber to which the experiment was proposed. They had a bigger chamber, an English chamber, and when the French said, "Look, it's silly, your chamber is not working, the beam's going, we've got the competition from America, why don't we take the chamber that is working and put in the beam?" And the answer was, "No, we can't do that because this experiment was proposed by this group for this chamber." So then the English said, "All right, we'll let you move your chamber into our beam only if the experiment remains an English experiment." So the French said, "No, it's ridiculous." So, there was a deadlock.

At CERN they have a policy where a certain group gets allocated a certain amount of time on the machine, i.e., two weeks. If they have trouble with their equipment, that's tough. If their equipment stops running on the last day, the next day it's off. These schedules are made way in advance and they are very difficult to change. But it has the advantage in that everybody knows when they are going to stop. They know that they are going to start on the fifth of October, and they start, because whoever is there before gets killed. Now this chamber, which I said didn't work, kept on being on actual location—it would work about two days from the end of each allocated period.

Now while they were being killed like this and having arguments about nationality, we had the laboratory guarantee that any time we could run, they would switch the full beam to us.

Any time we wished to have the full beam for any purpose, to find out what was wrong, they would give us the full beam. Any time we wished to go into the laboratory to do anything, to investigate our problems, they would turn the laboratory over to us. Everybody was standing back. Meanwhile they would go on doing other people's experiments as best they could, providing we had troubles, and this went on, I think, two and one-half months while we were desperately trying to make our beam work. Our chamber was working, our beam wasn't, and we worked day and night around the clock, three shifts, and eventually found out what was wrong. After we found out what was wrong, we fixed it. The accelerator went on full steam, and we got our film.

It's unheard of, in the rigid atmosphere of Europe. Somehow the administration here realized that this was a really interesting experiment, and it felt the competitive spirit too. It said, "Look, we're going to beat those CERN boys to it," and nobody complained about it. Everybody knew it was exciting; there was an atmosphere. CERN worked for a year; they're too straitlaced, everybody's interested in themselves.

He then tells about the outcome involving the politics of science and about some plans of his group that got interrupted.

Well, finally we found the Omega. We submitted that to the *Physical Review Letters,* and there is a ruling in *PRL* that you're not allowed to publish in any other paper, journal, or anything else, including newspapers, in advance of the publication date of *PRL.* So we called a news conference about a week before, with instructions that the newspapers weren't allowed to publish it. Actually, we got a concession that the *New York Times* could publish in the Sunday edition in spite of the fact that *PRL* was to appear on Monday, but they allowed that.

But there was a leak. The leak occurred through England, oddly enough. Someone in England knew by the grapevine that we had discovered the Omega and wrote a popular article for the *New Scientist,* which is an English journal like *Scientific American,* about SU_3, which he had written before any experimental results were available. He knew just before publication date that in fact we had found it, and he was tempted—and obviously fell to the temptation. He did not state that we had discovered it, but he did state that publication was imminent, which it was, but he didn't state publication by whom.

Then the game goes on. This was picked up by the London *New York Times* correspondent, and he knew that this was hot. He then interpreted it as the Omega *had been discovered* (and he didn't know who discovered it and couldn't find out who, but he knew who was looking for it—Brookhaven and CERN). So he said the particle had been discovered in Brookhaven and CERN, but the heading [byline] was London. There was no reference to the people who had worked on it here. There was a whole lot of splurge on the man who had written the article back in England who had nothing to do with either the theoretical or experimental discovery. It was a big mess. It killed our publicity. We would have got first-page *New York Times* Sunday, which is a very good thing to get. After all, where do we get national money?—from Congress. It makes a great deal of difference if we get first-page in the *New York Times* since most congressmen read the front page. It's a factor, and we can't deny it, so it matters an awful lot to us.

Then we kicked back. On Friday, we went to *PRL* and said they've released it, what do we do now? They said, "Okay, don't worry," and we backed up the date and told the *New York Times* that they could publish. What else could they do? So they published on Friday, stating really that it was us. We still didn't know if CERN had discovered it. We didn't know when it [the correspondent's report] said Brookhaven and CERN. We guessed that CERN was also about to publish. It took us until about Tuesday of the next week to discover that CERN hadn't discovered it at all, and the final news crept out that this was a pure Brookhaven discovery and we had done it first. By that time, it would have never hit the first page of anything.

British scientists in Britain told essentially the same story about the Omega-minus experiment, and there was a general consensus on the reasons why the Americans published first. A young experimentalist said:

It was definitely on the cards that we could have discovered it before Brookhaven. [What happened?] The British chamber, which we were taking pictures in, was late, and we used a smaller chamber for the time, and we didn't have the forces necessary to get the analysis done quickly. So the chamber lost us six months, I would say, and the delay in analysis probably lost us a year. There was chaotic organization for about a year in——— University. And the Americans published about a year before us. They were also subject to unforeseen circumstances and delay because their beam wasn't very good and things like that.

A more senior HEP offered this opinion:

The most striking example out of that film was the experiment on the Omega-minus that we started, but Brookhaven had their Omega-minus considerably before we had any examples out of ours. It was not only [the problems of our] large collaboration, it was partly due to the fact that they were rather better set up to do the analysis than we were.

A HEP not directly involved with the experiment himself was well aware of the progress of the two different rivals. He said with a rather sympathetic attitude:

There was a very good reason why we failed. [What was that?] Really lack of organization, I regret to say. They [the European collaboration] had the pictures—good pictures—but they couldn't do the analysis because they hadn't got a camera geometry complete, couldn't do a generalized reconstruction. This was due to the lack of liaison between the groups, pure and simple.

A detailed statement by a scientist who was probably more involved than the others quoted above gave a description that shows that the European collaboration almost beat the Americans in spite of their difficulties—even though the near-win resulted from chance.

There was one clear case of a race in the Omega-minus experiment. We had a head start, but technological problems arose—Brookhaven also had technical problems. We had already done one experiment that was an attempt to look for the Omega, but we knew it to be very much of a gamble. It was the best that could be done at the time with the technical facilities available.

We knew we had to change the experimental facilities to have a chance to find it, so at the same time we were doing the first experiment we set about designing the beam and getting everything ready to do the better experiment. So, not surprisingly, the Brookhaven people were set on doing the same part. We had a very good chance except the bubble chamber didn't work. So we got another, not a very good one, but we managed to convince the people at CERN that it was an important experiment and got them to drag the bubble chamber out of the other beam and stuff it in. We started taking pictures. One of the chaps who had been working on the beam, took a polaroid and said, "I've got an Omega!" Just an oddball event. But oddball in such a sense that it contained two of the characteristics we would be looking for as a signature of an Omega. To make this picture 100 percent Omega without doing any more except looking at it, we would have needed something like [here he described reconstruction of the physical event]. . . . If we had seen that then, we would have without question gone out and published straightaway. But that was missing. We took the film back and measured this thing about half-a-dozen times, and it was almost certainly an Omega. Just about the same time the Brookhaven people had started running, and they found their Omega. I got a phone call one morning from CERN saying thay had the Omega, so now we went back and said for sure we know it's an Omega. If the chamber had worked even half of the time, we would have been six months ahead. Of that experiment we did, we found nine Omegas, which is one-third to one-half of the world sample.

Priority conflicts result when scientists are involved in multiple discoveries, and of all types of creative endeavors in which individuals can engage scientific activity may be the one most likely to be involved in conflicts about who found or did it first. An example of this will be shown by the priority conflicts over the discovery of another new particle (the Omega-minus was also a new particle). This second particle will remain unlabeled to protect the identity of the claimants.

This case of disputed priority was reported by two British scientists. One scientist was working with an American group at the time of the dispute and the other was working in a European collaboration. The situation involves what appears from the British perspective to be an attempt of the Americans to claim a discovery first made elsewhere; from the American perspective, related by a British HEP, it appears to be a clear case of the American group's superiority. The scientist in the European collaboration at the time said:

We found a new elementary particle. It got around by word of mouth, and an American group quickly analyzed their data and tried to publish before us. [Did they succeed?] No, we found out about it, and one member of the group wrote to the editor of the *Physical Review Letters* and they stopped it. [How did the English group prove they had been onto it before?] Oh, it was just common knowledge. We *took some time* to analyze the data, to study it thoroughly, and in the meantime the other people here got the data and publication ready. They hadn't given us very full acknowledgment. They knew of our discovery beforehand and they hadn't given us full acknowledgment. They were somewhat dishonest. [Did you publish?] Yes, they didn't in fact get in before us. We were able to delay their publication so we got ours first [before they published] and they had to change the form of their reference. [You knew they were going to publish and had insufficient reference to your work?] Yes, we'd seen a preprint. Someone telephoned long distance and stopped them.

The time element in analyzing the research was substantiated by another scientist in Britain who commented on the same priority dispute:

If you ask me if the other people in this collaboration were embarrassed, the answer is yes. This was an instance in which we observed some new phenomenon and we told various people and the news very quickly spread to the States. Perhaps because we were *overconscientious or cautious* [in putting out our results] we had a preprint—a draft of a letter from an American group—also stating the same results. Some people were very upset because they felt these people had been given the lead to go and look at their data, because they had heard about our result prematurely. They claimed to see sinister signs in the way the method had been dealt with on the other side of the Atlantic.

The same incident was related somewhat differently by the British HEP who was working in America at the time. From his point of view, the question of priority was clear-cut, and in his group's favor.

There was a great deal of friction between the groups. There was some earlier relationship between the two groups in which we discussed what we were doing. And then we found that particle. At that very time a *personal* letter had been written by somebody in the [European] group who was working at———

to somebody in our group. As you understand, our group is bigger than the particular experiment[15] and so, in fact, that person was not on this experiment—he happened to be on vacation. So we went ahead and planned to publish. We didn't know about the letter because the guy had been on vacation, and in any case, he wasn't actually involved in the experiment, so the contact was very indirect.

So as a matter of courtesy, when we came to the point of actually being ready to publish, we wrote them a letter saying this is what we're going to publish, and got back telegrams and all sorts of hash about the fact that they thought that we were being very underhanded about that. They told us about what they had. You know when a thing is very close, like that, it depended on the functioning of a certain computer program, a matter of who got there first, and we got there first, and we got there, I thought, fairly clearly first. But in order to settle any quibble about it, because some people felt much more guilty about it than I did, we had a discussion in which we decided: shall we go along in what these people wanted to do—which was to publish simultaneously—or should we go our own free way? There was a vote about it. The result of the vote was to publish with them. So we published simultaneously, adjacent papers in the *Physical Review Letters*—ours first! That's what we agreed to since we had to hold ours back in fact to let theirs catch up.

It seems that in settling the dispute, some decision was made at a higher level than the two groups concerned. A senior scientist who was also involved in the priority dispute related how an eminent scientist helped settle the question.

. . . He had friends in America and he knew the editor. He rang him up and said, "Look here old boy, this is not the way you and I play. This is the story. Now you be a gentleman." In the end, the two letters were published side by side in the same issue.[16]

In most cases involving disputes of priority, there is the suggestion that the group (or person) who got there first actually did so—not because he was smarter or a better scientist—but because he had some unfair edge over his rival. This was illustrated when the same scientist said,

"Some people were very upset because they felt these people had been given the lead to go and look at their data, because they had heard about our result *prematurely.*" It may seem peculiar that any scientific information can be heard prematurely! To understand that, one must realize the nature of research, especially in bubble chamber physics from which this example came.

Bubble chamber research involves having thousands of black pictures with many white lines on them. Each line represents the path of some particle, and some of these are automatically interesting, some are not. Those not obviously interesting are potentially interesting if one knew what to look for. When two or more groups have film taken at similar energies with similar beams, it is likely that both groups could find the same thing *if they knew what to look for.* If one group unexpectedly comes up with another fruitful analysis, they realize that the other group could do the same analysis, and there is potential competition. A scientist told how that situation presently existed for him. His group and another group had similar film, but the others had quit work on their film. The others knew, however, that his group was still working on the film. He related how a scientist from another institution came to him recently and ". . . was very interested and had obviously been told by the people back home to find out what I'm doing, and I just didn't tell him. I prevaricated and missed appointments."

Premature information about what one is looking at can lead to being beaten, as this statement by a British scientist, working in the United States and involved in such an incident, shows:

We had so much film, CERN so much, and ——— University so much (less than us but more than CERN). We got our results out first, and reported it at the New York meeting, but we were a little bit slow in publishing it. So the week before we published, ——— University published a negative effect—said it's not there. But they didn't have enough film to tell if it

> was there or not there. They didn't have any degree of certainty whatever. In fact, the funny thing was that in almost the right place, they had a bump that anybody in their right mind would have to say, "Well, this is not negative evidence; it doesn't really add or subtract too much from what CERN had before, because we didn't have very much film." Really, the proper answer was, "I can't tell very much from this small quantity of film." But *because they knew that we had reported* the negative result with three times the film at the New York meetings, they interpreted their data in a heavily biased way, so that then, you see, the first published reference to this negative result is the ——— University result, because the New York meeting isn't published.

The experience has not been wasted on him: "Our policy now is not even to report at the meetings. Keep your mouth shut and then publish."

If a claimant in a disputed priority does not indicate that the other group got the word too early, they may accuse the group of unethical behavior.

> We had a certain result, which wasn't very significant statistically. ——— University had a result, which wasn't very statistically significant, and we got together at one time and talked about it. They said, "Should we publish it?" Then everybody agreed that there really wasn't enough data yet, and so they said, "Since we both have it and essentially discovered it during our conversation, if we ever publish, let's each tell the other we're going to publish before we do, so that we can publish simultaneously." So there was a verbal understanding that they would call us up before they published, and they just didn't do it. They just went ahead and published. So you just keep your mouth shut.

The race to find the Omega-minus and the case of disputed priority in the discovery of another new particle vividly illustrate the competitive nature of science in general and high energy physics in particular. Better organization in getting together at the right time all the variables affecting research is part of the explanation why one group accomplished its objectives before the other. The race for the Omega-minus suggests that British scientists experience anticipation quite often, but the disputed priority case shows that British scientists are able to compete and may not necessarily suffer being second all that often.

Competition in High Energy Physics

6

Prevalence of competition is measured by the number of times scientists have been anticipated through having research findings similar to their own published first by someone else. If they know their competitors, there may be a race, as happened in the Omega-minus episode. If competitors are unknown, a race is unlikely. It might be argued that running and losing a race is less severe for the person involved than losing out to an unknown person. At least one would know that he might lose, whereas not knowing one has competitors results in an unexpected and shocking loss.

Following Hagstrom (1966:1967), the HEP were asked if they had ever been anticipated. Most scientists, about two-thirds, reported that they had been anticipated (see table 11). Comparing that distribution to Hagstrom's sample of American scientists shows each nationality to have very similar experiences of anticipation (see table 12). While nearly the same proportion of British and American scientists have never been anticipated, if a scientist has ever been anticipated, an American is more likely to have been anticipated more frequently. Hagstrom (1967:3) reports that 2 percent of his sample had been anticipated six to ten times and 1 percent had been anticipated more

TABLE 11

EXPERIENCE OF ANTICIPATION BY HEP

Number of Anticipations	Percent
None	36
One	38
Two	17
Three	6
Four+	3
Total	100
(Base)	(202)*

*One HEP was not asked about his anticipation experiences.

TABLE 12

EXPERIENCE OF ANTICIPATION FOR AMERICAN SCIENTISTS, AMERICAN PHYSICISTS, AND BRITISH HEP

	I*	II**		III***	
Number of Anticipations	American Scientists	American Physicists Exper.	Theorist	British HEP Exper.	Theorist
None	37	40	36	35	37
1–2	46	46	43	57	51
3+	16	14	22	8	13
Total	99	100	101	100	101
(Base)	(1725)	(293)	(155)	(131)	(71)

*Source: Hagstrom (1967:3).
**Source: Hagstrom (1967:4).
***The chi-square for experimentalists and theorists is 4.36 (4 df., prob. <.50) based on the categories in Table 11. The mean number of anticipations for both experimentalists and theorists is 1.04.

than ten times. The largest number of times any HEP reported being anticipated was five times.

The differences between the two samples in the number of anticipations may be explained by the years of experience of each group. Among the British scientists, professional age is strongly related to the number of times they report having been anticipated (see table 13). Hagstrom also found a similar relationship. Controlling for professional age in the two samples, it turns out that the British

TABLE 13

EXPERIENCE OF ANTICIPATION, BY PROFESSIONAL AGE

Number of Anticipations	Professional Age* Younger	Older	Total
None	41	26	36
1	44	27	38
2	11	28	17
3	3	11	6
4+	1	8	3
Total	100	100	100
(*Base*)	(128)	(74)	(202)

$X^2=26.90$ (4 *df.*, *prob.* $<.001$)
gamma $=.45$

*HEP who are "younger" in professional age received their Ph.D.s after 1960. The "older" HEP received their Ph.D.s up until 1959.

TABLE 14

EXPERIENCE OF ANTICIPATION BY PROFESSIONAL AGE FOR AMERICAN SCIENTISTS* AND BRITISH HEP

	Professional Age**			
	Younger		Older	
Number of Anticipations	American	British HEP	American	British HEP
None	48	41	35	26
1–2	45	55	47	55
3+	8	4	18	19
Total	101	100	100	100
(*Base*)	(323)***	(128)	(1091)***	(74)

*Source: Derived from Table 1–3, Hagstrom (1967:10).
**Younger scientists in Hagstrom's sample received Ph.D.s in 1961 or later. Younger scientists in the British sample received their degrees in 1960 or later. Older scientists in both samples received their degrees in any year before the cut-off date for the younger scientists.
***The total for the American sample does not equal the total reported in Table 12 because these data were derived from a contingency table that required measurement on both variables.

scientists are slightly more likely to have experienced anticipation, but there are factors involved other than nationality that may account for the difference. The British scientists are all high energy physicists while the American sample includes scientists from many disciplines, and it could be assumed that high energy physics is a more competitive research area than the average of all areas in the United States. There is also a slight age difference in that the younger British scientists got their degrees a year earlier than the American sample (as presented in table 14), and thus have had another year in which they could experience anticipation.

Scientists who write many papers could reasonably be expected to experience anticipation more than scientists who produce few. That is, the higher the productivity, the greater the probability that at least one of the pieces of research will be anticipated by others. Hagstrom (1967:6) found this to be true among American scientists. For the British scientists, productivity is only slightly related to being anticipated (table not shown; chi-square significant at .50, gamma =.22). This difference is probably explained by the fact that the differences between the most productive and least productive British scientists are less than those differences between scientists in the United States. The lack of variability in British productivity would tend, therefore, to reduce the correlation.

It is reasonable to expect that scientists working on problems that they feel are among the most important

TABLE 15

EXPERIENCE OF ANTICIPATION BY WHETHER HEP WORKS IN ONE OF THREE MOST IMPORTANT PROBLEM AREAS

Number of Anticipations	HEP Works in One of Three Most Important Problem Areas: No	Yes	Total
None	42	33	36
1	37	37	37
2	18	18	18
3	2	8	6
4+	2	4	3
Total	101	100	100
(Base)	(62)	(134)	(196)

$X^2 = 4.97$ (4 *df.*, *prob.* $< .30$)
gamma = .21

TABLE 16

EXPERIENCE OF ANTICIPATION BY RANK*

Number of Anticipations	Rank: Lecturer	Sr. Lect./Reader	Professor	Total
None	40	24	26	36
1	43	24	21	38
2	13	22	42	17
3	3	19	5	6
4+	1	11	5	3
Total	100	100	99	100
(Base)	(146)	(37)	(19)	(202)
Mean Number of Anticipations	.83	1.68	1.42	1.04

$X^2=36.29$ (8 *df.*, *prob.* $<.001$)
gamma = .41

*Laboratory HEP were assigned academic ranks for purposes of analysis. Scientific officers and senior scientific officers were assigned "lecturer" rank. Professional scientific officers were assigned "Sr. Lecturer/Reader" rank. Senior professional scientific officers were assigned "Professor" rank.

for the future developments of high energy physics would be among those experiencing the most anticipations, since presumably they would be working in problem areas where there exists a discipline-wide consensus on importance. The data show this to be only slightly the case (see table 15). While it is the case that the HEP who reported their research problem area to be among the three most important areas are more likely than the others to have experienced anticipation, the percentage of each group experiencing one or two anticipations is exactly the same.

One question that can be raised—but unfortunately not answered—is whether the experience of anticipation has any effect on a scientist's career. If work is frequently anticipated, it might show an inability to successfully carry out research with dispatch. The data lend some support to the idea that being anticipated does affect one's chances of being a professor (see table 16). There could be other factors at work for which there is neither theory to suggest nor data to test them, so these observations are offered only to raise the issue.[1]

The largest percentage difference between the proportion of HEP who have never been anticipated results from the categorization according to rank. While 40 percent of the lecturers have never been anticipated, only 24 percent of the senior lecturer/reader rank have never been anticipated—a difference of 16 percent. (Professional age had a difference of 15 percent between the younger and older scientists.) Since professional age and rank are related and the number of cases is not large, it is impossible statistically to disentangle their relative importance.

There are two variables that show more differences in the experience of anticipation than any discussed so far. The first of these variables derives from the answer to the question of which scientists or groups of scientists have most influenced the scientists' current research. Answers were coded "other British scientists," "ourselves," "Europeans," "Americans," and so on. The raw distribution showed no differences between categories with the exception of those scientists currently influenced by American scientists, but that is a rather large and significant difference (see table 17). Only 17 percent of the scientists who said that their current research was influenced by American scientists have never been anticipated, while 43 percent have never been anticipated who said they have

TABLE 17

EXPERIENCE OF ANTICIPATION BY NATIONALITY OF SCIENTISTS WHO ARE MAIN INFLUENCE ON PRESENT RESEARCH

	Nationality of Scientists Who Are Main Influence on Present Research		
Number of Anticipations	American	British and European	Total
None	17	43	36
1	44	34	36
2+	39	23	27
Total	100	100	99
(Base)	(48)	(147)	(195)

$X^2=11.36$ (2 *df.*, *prob.* <.01)

been influenced by scientists located elsewhere. This result gives some evidence to the general observation that modern physics is stronger in the United States than elsewhere, and certainly for the last several years research in elementary particle physics has been centered in the United States. Scientists who follow the leaders and are influenced by them are simply more likely to experience anticipation.

The discussion of competition in the previous chapter, using the race for the Omega-minus and the priority dispute in the discovery of another new particle, indicated that communication between scientists at various stages of research can either help or hinder the competition. One would expect that modes of communication most used by HEP would be related to their experience of anticipation. If it is assumed that the mode of communication presently most important to a HEP has always been the most important, it is possible to suggest an explanation in addition to those above. HEP whose most important type of communication is verbal are the most likely to have been anticipated (see table 18). It seems they may have talked too much! And this relationship holds when controlling for professional age. Verbal communication

TABLE 18

EXPERIENCE OF ANTICIPATION BY MOST IMPORTANT METHOD OF OBTAINING INFORMATION RELEVANT TO RESEARCH

	Most Important Method of Obtaining Information Relevant to Research			
Number of Anticipations	Verbal	Conference/Seminar	Publications	Total
None	26	53	44	36
1	43	7	36	37
2+	30	40	20	26
Total	99	100	100	99
(Base)	(92)	(15)	(91)	(198)

$X^2 = 13.18$ (*4 df., prob.* $< .05$)

could be the main source of information, but that does not mean, of course, that HEP release information verbally.

Competition is severe to the extent that scientists lose the recognition that comes from doing research. If scientists were never able to publish as a result of being anticipated, there would be "complete" severity, whereas if scientists were always able to publish, the severity of anticipation (and the competition causing it) would be nonexistent. Although it is not possible to determine what is severe versus what is not severe, it is possible to compare the severity between American scientists and the HEP. As a result of the most recent anticipation, one out of five HEP did not publish. Hagstrom (1967:5 and 7) found that 33 percent of the American scientists were not able to publish, which indicates that their experience of anticipation is more severe than the British experience. Among the American scientists, 47 percent published because their work was different enough to warrant publication.[2] Five percent published because replication was desirable, 7 percent published for both reasons, and 8 percent published for other reasons.

Severity of competition for American scientists is related to the research area. For theoretical physicists, 50 percent were unable to publish; for experimental physicists 26 percent were unable to publish. Those differences are close approximations to the differences in severity experienced by British theorists and experimentalists (see table 19) except that more of both types of British HEP were able to publish.

Considering only the HEP, the differences in severity between older and younger scientists is of interest. Older scientists, more than younger, are able to publish as a result of anticipation (see table 20). Comparing tables 19 and 20, it can be seen that severity of competition is more a function of the type of HEP than of professional age. In table 19 the percentage difference in publishing after

TABLE 19

RESPONSE TO MOST RECENT EXPERIENCE OF ANTICIPATION, BY TYPE OF HEP

Response to Anticipation	Type of HEP: Experimentalist	Theorist	Total
Published	88	67	80
Did not publish	12	33	20
Total	100	100	100
(Base)	(83)	(45)	(128)

$X^2=8.41$ (1 *df.*, *prob.* <.01)

TABLE 20

RESPONSE TO MOST RECENT EXPERIENCE OF ANTICIPATION, PROFESSIONAL AGE, AND TYPE OF HEP

Response to Most Recent Experience of Anticipation by Professional Age

Response to Anticipation	Professional Age*: Younger	Older	Total
Published	75	89	80
Did not publish	25	11	20
Total	100	100	100
(Base)	(75)	(53)	(128)

$X^2=3.88$ (1 *df.*, *prob.* <.05)

Response to Most Recent Experience of Anticipation by Professional Age and Type of HEP

	Professional Age*				
	Younger		Older		
Response to	Type of HEP		Type of HEP		
Anticipation	Exp.	Theorist	Exp.	Theorist	Total
Published	82	63	95	73	80
Did not publish	18	37	5	27	20
Total	100	100	100	100	100
(Base)	(45)	(30)	(38)	(15)	(128)

$X^2=11.10$ (3 *df.*, *prob.* <.02)

*Younger HEP received their Ph.D.s in 1960 or later; older HEP received their Ph.D.s before 1960.

anticipation is 21 percent in favor of the experimentalist: in the upper part of table 20 the difference is 14 percent in favor of the older HEP, while in the lower part of table 20 the severity of competition is shown to be greatest for younger theorists, next for older theorists, next for younger experimentalists, and least for older experimentalists.

The "Matthew effect" is operative when considering recognition, productivity, and severity of competition.[3] Highly productive and highly recognized scientists, more than their colleagues categorized "low" on these variables, are able to publish after being anticipated (see table 21). The table for recognition is not shown because percentages in the two tables are identical. Why they are able to publish is not obvious. One hypothesis is that scientists whose work is better known are believed (by journal referees and other gatekeepers) to warrant publication even if it has been anticipated. Relatively "unknown" scientists may be viewed as trying to publish research simply because they are ignorant of the prior work. This hypothesis has some support from Hagstrom (1967:7) when he suggests that one of the reasons why 7 percent of his sample who published after anticipation did so possibly because they were unaware of earlier published work. But in this instance, they *did* publish, whereas in the Brit-

TABLE 21

RESPONSE TO MOST RECENT EXPERIENCE OF ANTICIPATION BY PRODUCTIVITY*

Response to Anticipation	Productivity Low	High	Total
Published	76	88	80
Did not publish	24	12	20
Total	100	100	100
(Base)	(79)	(49)	(128)

$X^2 = 2.68$ (1 *df.*, *prob.* <.20)

*The table for recognition is identical to this table.

ish case they did not. There may be different journal policies involved, and they are likely to be different journals since the different fields are not distinguished between those 7 percent.

An alternate hypothesis, and one that seems more plausible, is that young, low producers and scientists with low recognition may simply lack the confidence to submit their papers. A decision cannot be made between these hypotheses because there are no data on how many scientists submitted papers on anticipated research and subsequently had the papers rejected, although a young experimentalist provides some evidence for the latter hypothesis.

There's something I've worked on since I joined the bubble chamber group. It was a fairly theoretical problem that was interesting at the time. I worked it out and I was quite pleased with myself, and someone showed me a Berkeley preprint with it all written up. I was very disappointed about it. [He did not submit the paper.] It sort of struck me—the invincibility of Berkeley.

An area that neither Hagstrom nor I examined carefully involves whether or not a scientist distributed a preprint on work he did not publish because of anticipation. If a preprint was distributed, the recognition that would result had the person formally published[4] might still accrue to him. An example of this is given by a scientist who told of not publishing because of being anticipated.

I was beaten to the post by two other people whose papers I wasn't aware of. I was on sabbatical at ——— University [in the United States]. The other two papers that did much the same thing—one was about to appear in *Physical Review*—were not sent to where I was because ——— University didn't happen to be on the mailing list for a preprint for this kind of work. The other version appeared in *Nuovo Cimento*, and at that period *Nuovo Cimento* hadn't got across the Atlantic at the stage that I was writing my version of it. The paper was already out as a preprint, so I believe that in one or two subsequent publications *I got a reference* to my preprint, along with the people whose work actually got published. So, although it didn't get printed, *it got circulated.*

The prevalence of competition in Britain is very similar to that experienced by American scientists, although the severity of competition is less in Britain. The differences of the severity for theorists and experimentalists follow the pattern predicted first by Hagstrom (1965). The question is: Do British HEP react to being anticipated in the same ways that American scientists do?

When a competitive condition exists in science, scientists may deviate from usual norms when they feel such behavior may help them obtain the recognition they feel unable to receive otherwise. There are three main methods of deviance which scientists employ: (1) hasty publication; (2) fraud and theft; and (3) secrecy. Information about the first two types is difficult to obtain. One would meet with serious resistance if respondents were asked "Have you committed fraud?" The strategy was to obtain information from respondents about the possible deviancy of others. The third type of deviancy—secrecy—is relatively easy to discover.

Hagstrom (1967:8) asked his sample if hasty publication is a serious problem in their research specialty, and theoretical physicists said that it was. The HEP were not asked whether or not it was a problem, but they were asked whether journals had rejected any research they had submitted for publication. Some rejections occur because of attempts at hasty publication. Theorists, more often than experimentalists, have had their papers or letters rejected (see table 22). If journals rejected the paper, theorists were also often in disagreement with the editorial decision.

An instance of an attempt by experimentalists to publish hastily is given by a bubble chamber physicist. He said that the journal would not publish because "They said there wasn't sufficient analysis, but we thought it should be published for theorists to see if they could understand

TABLE 22

JOURNAL REJECTION OF SUBMITTED PAPER, BY TYPE OF HEP

Journal Rejection	Type of HEP Experimentalist	Theorist	Total
No	77	42	65
Yes—agreed with editorial comment	12	21	15
Yes—did not agree with editorial comment	12	37	20
Total	101	100	100
(Base)	(130)	(71)	(201)

X^2=25.52 (2 *df.*, *prob.* <.001)

it." Another experimentalist had tried to publish a letter, but found that "The editorial comment was that it wasn't of great enough urgency. I gave it to Professor ———, and he wrote saying it was, and it appeared." If the letter was initially believed to be a hasty publication without really being urgent, the editors obviously changed their minds when a person of some stature told them it was urgent.

Theorists often had letters rejected. Many were unhappy with the editorial decision. Hasty publication was clearly involved in some instances, but they did not want to admit this. One theorist tried to show that hasty publication was not involved:

This was a case when we did try to write a letter for we felt the work was not sufficiently large or *complete* to warrant a full-scale paper. We thought we'd publish some *preliminary* results in a letter because it was a topic of sufficient interest. But we discovered that *Physical Review Letters* had a peculiar policy that demanded a letter should be a finished paper with all the *complete* and final results. This is the only stand they took. There were a few other comments. We certainly didn't agree with the editorial comment. They had their rights, of course, as far as declaring their policy. But we didn't know of this policy before we submitted the letter so we felt it was ridiculous. They ought to have made this known to everybody and quite clear to everybody, but possibly they were going through a period of change of policy at the time.

This is an excellent example of double-talk. On the one hand, he was aware that letter journals are for publication of urgent information, but if other scientists cannot utilize the information the journal is not particularly interested. He also was aware that the work—meaning the problem attacked—was not too large. His intention may have been to publish a preliminary result that in all likelihood would have extinguished his whole problem. In the end, it seemed he forgave the journal for not "informing him," for if they were going through a policy change, *it was really no one's fault.*

Each case of having had a paper rejected was not as clear-cut as this one, but hasty publication obviously occurs. It will be instructive if future research is able to devise appropriate indicators of hasty publication. It may be that scientists who use hasty publication do not deviate in other ways, or it may be that deviants utilize all possible paths toward getting recognition when they are in a competitive situation.

HEP were not asked if they themselves had ever stolen an idea or if they had used any type of fraud to accomplish their objectives of publishing papers in order to get recognition. They did volunteer that other scientists had stolen their ideas or used some type of fraud.

Having ideas stolen seems to be a major threat to almost all HEP in the study. Experimentalists appear as anxious about having their ideas taken as theorists. Experimentalists probably are affected in ways different from theorists. The type of idea that the experimentalist is fearful of losing is *how* to do a certain experiment, not *which* experiment should be done. In contrast, theorists are more careful of *what* to do rather than *how* to do it. A theorist remarked:

I regard competitive activity as distinct from theft, which is starting with something when they hear someone has got it going. But very often all one needs to know is that there are things to be done in a certain area. *Anyone can do the things; finding the things to do is usually the problem* [my italics].

Interestingly enough, while experimentalists tend to be afraid of theft of ideas, not one was able to relate an instance in which his own idea was stolen. Of course, some suggested that in the process of bubble chamber analysis others had learned that a particular event might be found and they would then go and look for it, but that is not the same thing as having the idea stolen for a method of setting up an experiment.

In contrast to the lack of specific data from experimentalists, many theorists are able to recall having had ideas stolen. Even those who did not recall specific items are aware that the problem exists. When asked if he was aware of any instances of theft, a very young theorist said:

This is rather a difficult question, I know. Here you're getting down in the realms of muckraking, I'm afraid. It has been known that people will take or borrow or acquire other people's ideas. Alternatively, other people are afraid that their ideas will be stolen or borrowed. There is naturally in any group of people a desire to be first with an idea.

Theft of ideas may occur at any level of personal relationship. This happened to a young theorist who was working with a colleague, and as a result their relationship was strained, to say the least.

Basically what happened, I was doing some work at ——— University on a particular problem, and I did an enormous amount of calculations and decided at a certain stage that the particular method of calculation had reached sort of a dead end and that really it probably wasn't worth tackling, and so I stopped it. Now at about the same stage someone else in the department went to work at another institution for a while, on a sabbatical leave. And one day, he produced a paper on the identical subject with the same equations, as in fact I had been doing. What happened was, there was a small acknowledgment at the end of the paper saying thanks to [me] for some calculations, which basically boiled down to my having done the computations myself independently and him having sort of done exactly the same computations with the same theory that he got from the same source I had got it from—namely me! [Would you have considered it proper if he had made you a collaborator?] Yes.

Even one-time collaborators may become involved in questions of theft of ideas. One eminent theorist said,

Once I wrote a paper and someone else got all the credit for it. I had written a paper giving a certain method for finding a certain constant. Somebody I had already collaborated with got hold of data that wasn't available to me (I didn't know the data existed) and used it and wrote another paper. He got all the credit for getting the constant, and he didn't refer to me.

Young and relatively inexperienced theorists are not the only victims of such circumstances, as the last speaker shows. Another eminent theorist had a somewhat similar experience although the notion of a possible collaboration was not present. He said that

One person came to me and asked me for advice on a particular project in which he was engaged, and I spent quite a lot of time with this person and discussed various things with him. I was also engaged in similar projects, and I told him what I was doing—how far I'd got. And the next thing I knew he had published essentially my work. He had memorized enough of it and gone off and written something quickly.

There may be some problem in interpreting the theft of ideas when scientists are closely associated at some point in time. Scientists exposed to the same seminars, coffee time discussions, and so on, are likely to suffer from *cryptomnesia* (see Merton 1963) or unconscious plagiarism. Whether or not cryptomnesia is what happens or whether conscious theft of ideas is what really happens probably depends upon the source of information as much as the actual event. There is, of course, the tendency in an interview situation to believe that the informant is the innocent victim while the unknown "other" is the guilty party.

Theft of ideas may occur at a different level from interpersonal contact. Without personal contact, "theft" may occur as a result of the use of published or unpublished ideas as though they are one's own without reference to the person who had the idea and outright theft through devious means. That having one's work cited is a source of recognition and pride is hardly debatable. One experimentalist said when reporting that he had not been cited enough: "Some of them were my friends and I told them off." Even among the most eminent scientists, when they already have wide reputations, there is a sensitivity about being cited. One experimentalist told what to him was an amusing story.

I remember giving a review talk at the ——— Laboratory. X and Y were buddies at one time. Well, I mentioned in this talk that work was done by X et al. I didn't include Y. Y was there, and when it was over he took me aside and told me I should refer to him also. I kind of laughed, but he was quite serious and took it to heart.

HEP were asked if other scientists had failed to cite their papers when it was clearly called for. Half of the scientists had no knowledge of this having happened to them.[5] For the half who did have this complaint, 43 percent knew the people involved; 20 percent both knew the scientists and believed that the lack of citation to their work was intentional.

Other scientists were excused from not citing their work for reasons such as "no one can keep up with the literature," "it was published in a journal that gets to him late," and so on. Europeans may be in a particularly difficult situation in getting their work cited as often as it should be.

Remember that when you submit something to *Nuovo Cimento* they write back and accept it. You then have to wait a year at least before it comes out. If during that time someone anticipates your work and writes to *Physical Review Letters,* then in fact they would beat you to publication. This is life. And as such, of course, they won't refer to your paper, and

> this is the sort of lack of communication I mean. In general, you have got to have all of high energy physics in preprints to avoid this problem.

And it has already been noted that many institutions do not receive preprints, which puts scientists working at these institutions at a disadvantage (Libbey and Zaltman, 1967).

Among those who believe that failure to cite is intentional but the person is not known personally, respondents obviously have to do some guessing about whether or not the failure to cite was intentional. The problem of the allegation's validity is not the concern here. What is important is what the scientist believes. When asked to explain why scientists have not referred to his work, a theorist replied,

> Possibly a reference to my work would distract from the importance of their work, because what I've done so far in my field has always been new. The approach has always been better than what was done before and what has happened since. There's been very little improvement.

A similar sentiment was expressed by a more eminent theorist: "It very often happens that people who haven't published much will not refer to your work because the only way they can get their paper into print is by not referring to the preceding paper that has done the same thing." In such cases an intentional refusal to cite can result from simultaneous discovery, in which case the lack of citation does not imply use of another scientist's ideas in any way. It is still an attempt to ignore other research. One theorist said:

> A group of three of us were doing some thinking, talking among ourselves about some idea at CERN, and we made a discovery. Then publication came to some Americans who had the same idea. Therefore, we got working fast to publish ours. We didn't refer to the Americans. We had a slightly different method.

It might have been better in the long run to have referred

to the others, because the letter was rejected by the journal on grounds of nonurgency.

It is not surprising to hear that theorists believe others refuse to refer to their own "superior" research because, as Hagstrom (1965) suggests, theoretical work does not require as much replication. The data on Hagstrom's (1967) American scientists and the HEP uphold this hypothesis: theorists publish less often than experimentalists after being anticipated. It is somewhat surprising to find that experimental scientists also impute similar motives to other experimentalists; namely, that referring to other research distracts from their own.

> I suspect whatever they published and actually made a large part of their publication was stuff that we had developed and published. If they had given our reference it would have appeared very obvious that what they had done was something that had already been done.

This type of remark is especially surprising when considering the frequent statements experimentalists make that "replication is desirable." There is double-talk involved in statements of this kind. Experimentalists say they want to be first, but when they are anticipated they report they publish anyway because "replication is good." Rationalization for being second is naturally a factor. For example, one scientist said, "We found several things they didn't. Ours was a good job. Sometimes duplication is a good thing." Another commented, "Duplication in experiments is not a bad thing. *Of course,* you want the first major reference on a topic." When asked "Had you got your results before they did, would you be more satisfied?" a scientist replied, "Of course. Extremely rarely is the confirmation recognized as much." A more experienced experimentalist confirmed this opinion: "Unless you manage to get your second paper well established, that is, presented at important conferences, which is probably the easiest way, people don't bother to check up on what further work has been done."

Another kind of rationalization may be employed when scientists argue that "To be first doesn't always mean that you are right. You may publish what you do, and then later someone may do a follow-up and find that your results were wrong and come up with a better answer. So now everyone is always trying to be first, but the one who is right is what matters." That being right matters for science may be correct, but what matters for the individual is shown in a statement that shows a conflict between what is good for science and good for the individual. "It's the first man who gets the credit. [He will be referred to?] That's right. Even in some cases where the original experiment has been considerably wrong, it's always referred to."

The scientists who both knew the people who did not cite their work and knew the lack of reference was intentional have concrete evidence for their beliefs. They have had more experience and know the habits of the scientists involved. One experimentalist reported that

> We sent a preprint to ——— and in a matter of weeks received a preprint from them showing the same effect, but in a different reaction that didn't give a reference to ours. They also had sent the letter to *Physical Review Letters.* Ours had been delayed at preprint stage before being sent to *Physical Review Letters,* so they arrived about the same time. It was very complicated. We phoned the editor in New York and in the end got ——— to make a reference to our work.

Personality differences and clashes are often involved in not citing other scientists' work. One experimentalist said, "X [an American] doesn't recognize any work but his own." Another experimentalist replied to the question of why some scientists did not refer to his work:

> Idiosyncrasy of the personality involved. It was while I was at CERN doing an experiment. We published the data, and it was later done by a group at ———, more precise and a better experiment than ours. But the leader of the group just considered that our work wasn't worth talking about. The group leader was Y [an American].

The discussion to this point has focused only on the victims of others' failure to refer. One scientist reported that he was personally guilty but when he tried to amend his wrongdoing the injured party would not accept. The injured theorist now refused to refer to him.

> The first paper I published was remarkably similar in nature to one someone else had done. He wrote to me pointing out that he had already published it. The technique we used was the same, but he had hidden it in ———, which you usually don't take a look at. Furthermore, he had put it in the middle of a very long article. Consequently, two other people [journal referees] and I overlooked it. He [the injured theorist] continues to publish without reference to my article because he can refer to his, although they're not exactly the same. I rather think it would be politer if he at least acknowledged the presence of mine. I certainly acknowledge the presence of his. But his work is well known. He's X and at ——— University. He's a ——— [rank] now. He ought to be a professor but isn't because he's so annoying.

There are formal and informal sanctions for failure to refer to other research. The informal sanctions involve those scientists who earn a reputation for non-reference, a reputation that travels by word of mouth. Distribution of preprints provides a channel of communication without referees. Any control over nonreference is missing and there is no immediate sanction such as requirements by editors to cite others. References to previous work assist to allocate recognition, so the most important sanction, in addition to spreading the word about scientists who habitually refuse to refer to previous research, is to treat them in the same way, as illustrated in the above quotation.

The formal sanction was also previously illustrated. If referees, editors, or the injured party know a paper has not referred to the relevant research, the first two may require it before publication while the latter may lodge a complaint to the editor and cause the publication to carry proper citation.

In contrast to theft of ideas when the persons involved

are known to each other either personally or through second parties (or reputation), there are instances of theft that involve an obvious intent to take the other person's work and pass it off as one's own. It should be noted that this type of deviant behavior probably occurs very seldom, but the fact that it does occur at all is interesting. One case involved a student-professor relationship. A theorist related this story of theft:

> At the time I was a student at ——— University [in the United States], working with another student. We had finished the paper and we gave it to the faculty. They subtly suggested we not publish. In fact, one of the faculty published it who had been asked to give his criticism. It was a dirty deal.

Additionally, two scientists were concerned about theft by journal referees, and another had experienced actual theft.[6] The three cases form a continuum based on the degree to which one can put trust in the validity of their remarks. The first scientist talked about what he suspected about referees.

> You must realize that the ethics usual in the scholarly pursuits, literary and so on, do not apply to high energy physics. Outright dishonesty is prevalent, and there's not that much stigma attached to being caught at it. I suppose the referee system is not too bad, but it also is definitely used immorally—to delay your competitor. If you happen to be a referee for a particular article, it could frequently be policy to delay—I have not been in this position on either side, but it definitely does happen. I've also used considerations of this sort [to determine] to which journal I send the paper. Very often one has a very reasonable idea of who is liable to be a referee. For example, if I were to send a paper to ——— [journal], which I thought was going to a certain person in ——— [university], for example, who might well want to delay my paper, then I would probably send the paper to ——— [a different journal].

The system that he describes certainly is not that bad, for if it were no one could ever publish. Presumably journal referees are both knowledgeable and are doing similar research themselves at the time they are called on for judgment. In that case, every paper submitted is read by someone who is a potential competitor. The reason for believing that referees are corrupt may result from some knowledge of instances like the following. A young scientist, having more than mere suspicion, described how a paper he had sent in had been rejected because the referee said the paper was incorrect.

The results were subsequently used by the referee. The work was published later. I won't give you his name, but at that time I was pretty green. I derived my result and submitted it, and the referee's comments were to the effect that the calculation was incorrect and invalid. Then six months later it was in print. [How did you know who the referee was?] I saw it later. *Nuovo Cimento* has European referees, mostly from CERN, and there were only three or four in that field at that time, so I deduced it after seeing the paper, knowing all the time that the paper was correct.

The ultimate theft would be to steal a preprint, change the author's name and institution, and submit it to a journal as one's own work. All the credit would be lost to the victimized scientist. If a paper has *already* been published and is then stolen, very little credit may be lost. If it is published in a journal that "important" physicists read, it will be recognized as already published. When one scientist was asked if another scientist had failed to refer to his research, he said:

No. There was a case, though, where one of my papers was published directly. A friend of mine found it in the ——— [journal]. The paper I had written was copied and published under somebody else's name. [How did they get hold of it?] Presumably from a preprint—I don't know—or from my journal article that appeared a long time after the preprint. Even the graphs—he just copied the graphs. I wrote asking for a reprint but never got any answers.[7]

Secrecy is the most prevalent type of deviancy that scientists employ as a response to competition. Hagstrom (1967:14) states that "The scientist in a competitive situation will tend not to disclose his ideas to his colleagues until he is ready to publish an article that will assure him of recognition. He may fear that others will use his ideas to solve his research problems before he does so himself. . . ." Hagstrom asked his large sample of American scientists the following question: "Would you feel quite safe in discussing your current research with other persons doing similar work in other institutions or do you think it necessary to conceal the details of your work from some of them until you are ready to publish?" Among the physicists who responded, 42 percent of the theorists and 51 percent of the experimentalists would not feel safe in discussing their work with all others (Hagstrom, 1967:126). Theorists are less hesitant to discuss their work than experimentalists, but the point should be noted that the question referred to current research and not to research in general. Hagstrom (1967:15, 17) suggests that secretive behavior probably reduces the rate of discovery and also takes some of the pleasure out of doing research because informal recognition is denied at the early stages of research.

The HEP were asked a somewhat different question about discussing their work: "In some cases, people may feel it advisable to refrain from discussing their work with certain individuals or groups, especially if they are at certain stages of their research. Is there anyone with whom you would *not* discuss your work?" This opened the opportunity to get data on types of secretive behavior and at what stage the secrecy was likely to be felt most important to the individual. Of all the HEP, 41 percent indicated that they would always discuss their work and, as Hagstrom found among American physicists, more experimentalists than theorists are secretive (see table 23).

Hagstrom (1967:16) found that secrecy is highly correlated with concern about being anticipated. Among all scientists in his sample, only 32 percent of those *not at all* concerned about being anticipated in their current work are secretive, but 77 percent of those already antici-

TABLE 23

SECRETIVE BEHAVIOR*, BY TYPE OF HEP

Secretive Behavior	Type of HEP Experimentalist	Theorist	Total
No	37	49	41
Yes	63	51	59
Total	100	100	100
(Base)	(132)	(71)	(203)

X^2=2.82 (1 *df.*, *prob.* <.10)

*Secretive behavior is based on the HEP answer to the question whether he would refrain from discussing his work with some other scientist at any stage of research.

pated (or very concerned about being anticipated) are secretive. It is important to note that 32 percent of the scientists not concerned about anticipation are secretive. One explanation for this is that they could be secretive because there is a high probability that they will be anticipated but it is of no great concern to them, although they are taking precautionary steps. Another explanation is that secrecy is also related to factors other than concern about anticipation. This hypothesis will be considered again.

Hagstrom does not suggest that there is a one-to-one relationship between concern about anticipation and secrecy; nor does he suggest that competition always leads to secrecy. But the general model is that competition causes scientists to be anticipated and concern about being anticipated causes scientists to be secretive. This seems to be a straightforward model, but there may be complicated processes that are covered up when considering only these variables. Among both American and British scientists, theorists more than experimentalists lose recognition because of failure to publish, which means that competition is more severe for theorists. But experimentalists are more secretive (as measured by different questions in each sample). For the British, severity of competition—whether or not publication was possible after being anticipated—is not related to secrecy. Competition, with the experience of anticipation as an indicator, is prevalent among British HEP. To see the extent to which competition fosters secrecy, it is necessary to look further than the raw percentages that showed that experimentalists are more secretive than theorists.

If concern about being anticipated is related to secrecy, it could be argued that secrecy is related to the probability that present research may be anticipated. If being secretive about one's research provides a competitive advantage, then research that is likely to be anticipated would not be discussed in order to protect the small possibility that it might not be anticipated. The data do not uphold that hypothesis, showing instead that for *all* HEP there is no such relationship (see table 24). If we consider type of HEP, there is an intriguing relationship (see table 25). Increased probability of anticipation has little effect on experimentalists. The reason experimentalists appear to be consistently secretive will be shown later. If the probability is low that research is going to be anticipated, theorists are secretive, but if the probability is high, they are not secretive. In other words, secrecy is negatively related

TABLE 24

SECRETIVE BEHAVIOR*, BY PROBABILITY OF PRESENT RESEARCH BEING ANTICIPATED

Secretive Behavior	Probability of Present Research Being Anticipated: Low	Medium	High	Total
No	36	38	46	40
Yes	64	62	54	60
Total	100	100	100	100
(Base)	(99)	(26)	(68)	(193)

$X^2 = 1.46$ (2 *df.*, *prob.* $< .30$)
gamma = − .07

*Secretive behavior is based on the HEP answer to the question whether he would refrain from discussing his work with some other scientist at any stage of the research.

TABLE 25

SECRETIVE BEHAVIOR*, PROBABILITY OF PRESENT RESEARCH BEING ANTICIPATED, BY TYPE OF HEP

Secretive Behavior	Type of HEP						
	Experimentalists			Theorists			
	Probability of Present Research being Anticipated			Probability of Present Research Being Anticipated			
	Low	Medium	High	Low	Medium	High	Total
No	38	27	36	33	55	67	40
Yes	62	73	64	67	45	33	60
Total	100	100	100	100	100	100	100
(Base)	(63)	(15)	(47)	(36)	(11)	(21)	(193)

X^2 (*Experimentalists*) = 0.687 (2 *df.*, *prob.* <.50); *gamma* = .05
X^2 (*Theorists*) = 6.21 (2 *df.*, *prob.* <.05); *gamma* = −.50
X^2 (*Low*) = 0.224 (1 *df.*, *prob.* <.70)
X^2 (*Medium*) = 2.08 (1 *df.*, *prob.* <.20)
X^2 (*High*) = 5.44 (1 *df.*, *prob.* <.02)
X^2 (*Table*) = 9.36 (5 *df.*, *prob.* <.10)

*Secretive behavior is based on the HEP answer to the question whether he would refrain from discussing his work with some other scientist at any stage of the research.

to probability of anticipation for theorists. The explanation for this is the nature of theoretical work. Theorists indicate that knowing which research problem to work on is the most important aspect of their research. When theorists are secretive but the probability of their work being anticipated is low, it is necessary to look further to see why they say the probability is low. The explanation for their secretive behavior is that theorists want to maintain the low probability of anticipation. This is shown by the fact that 75 percent of the thirty-six theorists who say the probability of anticipation is low indicate that it is low because no one else is working on their problem. They have found a problem that they think is their own, and they intend to keep it that way! This suggests that—at least among British HEP—secrecy is not simply a function of the probability of being anticipated. These findings are somewhat at odds with Hagstrom's (1967) results. It can be argued that concern about anticipation, which Hagstrom related to secrecy, is not the same as relating the probability of being anticipated to secrecy. These variables are probably interrelated since both are related to a third variable, the experience of anticipation.[8] The question of which measure—concern about or probability of anticipation—is more appropriate cannot be answered here. One can only suggest that future research should take cognizance of the problem.

It could be argued that increased probability of anticipation is not related to secrecy in a positive way for the HEP because the British experience of anticipation and the utility of secrecy have a meaning for them different from that for American scientists. That could include the different possibilities for tenure, promotion, and mobility in Britain and the United States, but these differences would not explain the data because the British appear to be as secretive as the American scientists.

There is another possibility, therefore, which should be considered. As already noted, some American scientists who are not concerned about anticipation are nevertheless secretive and some scientists who are very concerned are not secretive. Some British scientists likewise do not conform to the model of secretive behavior. This means that at least part of the explanation for secrecy must lie elsewhere. Responses by the HEP indicate that some secrecy is learned simply in the process of becoming a scientist and may not be related at all to present competitive conditions. Here reticence would be a better term than secrecy.

In addition to information about whether the HEP would discuss their work, they were asked what types of information they would withhold. There is a striking difference between experimentalists and theorists (see table 26). Over half of the secretive experimentalists would not discuss their work after they had done only preliminary analysis of the data. The same proportion of theorists would not discuss their work at the idea stage. Less than one-fifth of each group indicated their secrecy extended to the time of publication, and so secretive behavior is often used for reasons other than keeping competitors at a disadvantage.

TABLE 26

STAGE OF RESEARCH OF SECRETIVE BEHAVIOR*, BY TYPE OF HEP

Stage of Research of Secretive Behavior	Type of HEP Experimentalist	Theorist	Total
Idea stage	20	53	30
During experiment/working out theoretical problem	12	25	16
After preliminary analysis	53	3	38
Before publication	14	19	16
Total	99	100	100
(Base)	(83)	(36)	(119)

X^2=28.44 (3 *df.*, *prob.* <.001)

*Secretive behavior is based on the HEP answer to the question whether he would refrain from discussing his work with some other scientist at any stage of the research.

Some experimentalists were afraid of discussion at the idea stage. One remarked that

> If one did have a new idea, and one was thinking of proposing an experiment to do it, it's unlikely that you would go around broadcasting this idea, because there are so many accelerators there's nothing to stop anybody else immediately taking an idea and proposing an experiment.

Theorists more than experimentalists are secretive at the idea stage. An eminent theorist said:

> There are many people who are not very much aware of who they get ideas from, and six months after they hear the idea they forget who suggested it. Talking to those people about your ideas can get yourself into trouble.

Not all theorists who are secretive at the idea stage are afraid of having the idea stolen. Others are concerned about their reputation in proposing "stupid" ideas. An experienced but reticent theorist related:

> I think this depends very much on the *personality* and attitude of the person you may or may not discuss ideas with. I mean, there are certain stages in my work where I might be inhibited about discussing it with certain people because I know that I haven't got my ideas fully worked out, in which case there are some people I might discuss it with who will simply show me what a fool I am. This happened in early discussion with one or two of my colleagues, who shall be nameless.

Although this same feeling was expressed by the experimentalist in Wilson's novel (1949:104) when he said: "Everybody says go out and think for yourself, but if you do and come back with a new idea they damn near kill you. So you can't be afraid of what people will say and you can't be afraid of your own judgment." This bold character could make that adjustment since he was hiding in the pages of fiction.

Theorists are likely to be secretive while working out a problem to avoid giving competitors an edge. A young theorist suggested that "If one thinks one is at the stage of hatching out something rather good, one might not want the other person to know. One wouldn't want them to hatch it out as well."

Experimentalists who are secretive during the process of an experiment may have different reasons for not discussing their work with colleagues. One of these reasons is to increase their competitive position, of course. Describing when he had recently been secretive, one experimentalist remarked:

> Well, the situation was one in which we were doing an experiment that another group was setting up to do. It was feared that if they were aware of how far ahead we had gone with that experiment they would have pushed a lot harder than they were pushing. They knew what we were doing, they knew there was an overlap between what we were doing, but we certainly did conceal from them exact information of just where we were on that experiment.

As a matter of policy some scientists refrain from discussion of certain aspects of their experiments. An experimentalist based his type of secretive behavior on laboratory observations.

> If an experiment is not going well, it is not advisable to "shout it from the housetops" because it might get thrown off the accelerator. There's an experiment at the moment just creeping up from the ashes where things were going very badly experimentally. They weren't, in fact, observing the events they had proposed to see and measure, and they very nearly got turned out of the experimental area before the experiment was started—let alone before it was finished.

The problem of doing experiments and the subsequent reputation that scientists may get as a result of discussing certain aspects of their research are illustrated by a senior experimentalist:

> In most experiments, because you're always trying to do things that have never been done before, you are usually—if an experiment is really a worthwhile experiment—*just about* successful. If people are honest, they'd agree that they are not as successful as they had originally planned. So there is a stage of the experiment at which I would prefer not to talk until one has

sorted out the problem for oneself, because you very often go through a stage where it looks as though you have terrible problems. If you discuss these—you know the way word gets around the laboratory that (a) so-and-so is having a hell of a job deciding how many particles he has in his beam and (b) they don't understand so-and-so. And, if you're not careful that becomes a part of the experiment. When it is published or discussed, people say, "Oh, that was the experiment where such-and-such happened, I don't believe that," when it was no different from any other experiment.

Theorists rarely indicate secretive behavior after preliminary analysis, but this is when experimentalists tend to be secretive. One obvious reason for not discussing experimental results involves giving out wrong figures. This type of secretive behavior is probably efficient for science. Some HEP indicated they were secretive because of the harm to science that otherwise would result:

There would be occasions when I would not be prepared to give anyone numbers to enable him to interpret the experiment when I am myself not certain yet that those numbers are true numbers. But I would always be prepared to discuss qualitatively what's happened. I would show him our tentative numbers and say what interpretations I have put on them, but I would not wish him to take them away and do something with them himself until I'm sure that the numbers are good numbers.

Another experimentalist told about premature discussion of an experimental finding that had implications for his group's reputation as scientists:

We had a rather embarrassing episode with the ——— University group. We collaborated with them on an experiment. In the early stages of the experiment, we found an interesting event that the ——— University people got very excited about—which I didn't think was a very good event. Anyway, "X" blabbed this to "Y," and this got around in the States. "Y" told around that we had found this interesting event, and we got another enormous exposure at the ——— laboratory. It turned out that the one event just wasn't a good event. Well, you know, if we found the same thing again, we wouldn't be so quick to discuss it like that.

A younger experimentalist said, "I would not give out preliminary results before I'm satisfied that they were right to prevent one from getting a reputation for 'jumping on the bandwagon' and releasing numbers 'hot from the press' when you haven't got time to consider them." A reputation of inability to interpret one's own research correctly is a problem when people are anxious to know the "number" that resulted from an experiment. An experimentalist said that

> If you have some result that is tentative, you obviously don't want to speak about it if you're not sure; you don't want to make a fool of yourself. I know that whilst we're supposedly grown men, you get childish rivalries coming in. Physicists are certainly a human lot, there's no doubt about that. They can even be a ruthless lot. I think you will find as much of this here as in industry.

Secrecy until publication of results is probably the most harmful for scientific progress. The scientific community must wait until every aspect of the research is published before it can use the results to guide new research or correct an earlier perspective.

An eminent theorist suggested the competitive reason for secrecy until publication and also the reason involving personal recognition.

> One does have certain rivals to some extent who are doing very similar work. If you get something that is very new, very surprising, you don't like to dish this information to them on a plate. You like to wait until a conference comes along and then dish it up to them. It's just very nice from your point if you're able to stand up and claim that you're the first person who has seen this event or got this number or shown that so-and-so was wrong. It's just this: it's much nicer being in front of five hundred to one thousand people than writing it on a piece of paper and submitting it to a journal. Purely psychological thing!

One of the most persistent complaints among experimental HEP is the fact that theorists—especially phenomenologists—are anxious to get the data to feed into their theoretical models. Scientists, whether they are secretive or not, want to publish their own data first. Some are thus secretive until they can publish it because

> In general terms, there are a large number of theoreticians who hang around like vultures, waiting for data to come out of the machine. And you find—we have found this in the past—that something gets published involving your data before you even get time to publish.

A theorist was very sympathetic about the plight of the experimentalists. When asked whether or not he has requested such preliminary data, he replied that

> Experimentalists are feeling rather bad about this. They take two to three years to produce an awful lot of data, and they just produce the data and that's the end of it. Then, a group of theoreticians get hold of the data and analyze it and produce a whole set of resonances [particles]. People take far more note of resonances than they do of just blocks of data. The theoreticians that analyze this data become far more widely read and widely known than the experimentalist who has produced the data—and the experimentalists don't like this.

It seems rather ironic that theorists can become more widely known using others' data, but this supports the conclusion that theorists receive more recognition from the scientific community than experimentalists. Apparently this problem of experimentalists feeling possessive about their data (at least until it is published) has been in science for as long as role differentiation has been present. Mitchell Wilson (1949:156) must have observed this conflict before he included this scene in his novel, *Live with Lightning:*

> "As far as Dr. Carter goes, I am the last man to criticize the presence of her sex in a man's profession," said [Professor] Regan with an attempt at Old-world charm. "The last man, may I add, only because I am extremely polite and will wait my turn at the end of a long, long line of men who have not finished with their protests. . . . But her presence here is symptomatic of a larger degeneration. This paper, for example, was concerned with a mathematical approach to experimental science. Why,

that's pure gibberish. The life of science is experiment, experiment, experiment. Who is this newfound theoretical physicist? Not Dr. Carter, *per se,* but the whole damn kit and caboodle that has jumped on our backs like a bunch of peddlers, suddenly turned shopkeepers. And what do they keep shop with? What's their merchandise now? Why nothing but the hashed-over, rewarmed, sauced-up, sliced-down, God-given results of the experimentalist! All dressed up in the gibberish of the so-called mathematics, which if we were honest, not one of us really understands!"

One experimentalist expressed his sentiments very explicitly in speaking of the problem of theorists using experimentalists' data, but he added a rationale that probably enables life for the scientist to go on in a rather peaceful and congenial manner. "Theorists are bad," he said, "about taking unpublished data and trying to use it to form their theories. Many times it is not sufficiently analyzed. Again, physicists are people. Some sons-of-a-bitch are excellent physicists. You tolerate them for that reason. If they were mediocre, you wouldn't tolerate them."

These statements show that scientists are secretive at various stages of research for various reasons. HEP concerned about their reputation are secretive during the preliminary analysis of data (see table 27). HEP who indicate secretive behavior because of their desire to be first are secretive at various stages of their research. Relatively few HEP are secretive because of a personal policy that they have developed, but for those few the secretive behavior is at the idea stage of research.

Science is competitive, and competition sometimes leads to some secrecy, but some secretiveness results from young scientists' role models. For example, one scientist said, "I've been advised—this was when I was doing the Ph.D.—not to speak too much about the idea of doing an experiment in case somebody else gets the idea." His statement does not suggest lack of competition but rather an effect that resulted from competition present at an

TABLE 27

STAGE OF RESEARCH OF SECRETIVE BEHAVIOR*, BY REASON FOR SECRETIVE BEHAVIOR

Stage of Research of Secretive Behavior	Reason for Secretive Behavior			
	Personal Reputation	Desire for Scientific "First"	Personal Policy (Practice)	Total
Idea stage	11	34	59	31
During experiment/working out theoretical problem	17	15	24	17
After preliminary analysis	72	20	18	36
Before publication	0	31	0	17
Total	100	100	101	101
(Base)	(36)	(61)	(17)	(114)

$X^2 = 45.52$ (6 *df.*, *prob.* $<.001$)

*Secretive behavior is based on the HEP answer to the question whether he would refrain from discussing his work with some other scientist at any stage of the research.

earlier time that may or may not be present at the moment.

In various fields of science, especially those in which the time required to complete experiments is considerably shorter than the two-to-three years required in HEP, it may be the case that different patterns of secrecy are prevalent. It could be argued that secrecy because of competition would be more prevalent where results could be published in a short time, as with experiments with relatively simple apparatus. In those sciences, different factors might be operative that affect the prevalence of secrecy. Whatever the causes of secretiveness in a given area, competition for recognition is one of the more interesting dilemmas facing contemporary scientists.[9]

Communication in the High Energy Physics Community 7

Communication is the lifeblood of science. If a communication system in science is efficient, there is no duplication of unnecessary effort. Time and other resources that might be wasted on such efforts are utilized elsewhere.

The communication system in science affects the social organization of science. If the communication system is relatively inefficient so that information about the progress of everyone's work is not known to everyone else, some scientists develop a social organization to insure themselves that they—at least—are informed. The social organization of science also affects the communication system. If a specific field contains many scientists located in a great number of different institutions, one type of communication system may be required; if there are relatively few scientists in a few different institutions, another type of system may be necessary. The point is that the communication system and the social organization of science are interdependent.

If a totally efficient communication system existed, everyone would know about each other's work. The opposite extreme would be where no one knew anything about anyone else's work. The reality of the situation is obviously somewhere in between these extremes, and is produced by two channels of communication—formal and informal.

Formal channels of communication are adequate for the distribution of necessary information if immediate publication of research occurs when the communication is received and the growth of this particular scientific field is rather slow. The first condition is never met, and the second condition is rather rare, at least in the natural sciences. Informal channels of communication thus become important aspects in the total communication system.

One main purpose of informal communication is to circumvent the slower formal channels in an attempt to establish priority or stake claims to an area. The best example of this is the widespread use of preprints that when distributed by laboratories, agencies, or large university departments assume the characteristics of formal publication.[1] Another purpose of informal communication is to assist a scientist's work by helping to plan future research and to revise present work.[2] Two exploratory studies of the communicative behavior of scientists show the importance of this function of informal communication. Glass and Norwood (1959:196) found that casual conversation was the most important method of communication that fifty scientists used to keep up with the progress in their own fields. The second most important method was regularly scanning journals. Menzel (1962:423) shows that accidentally obtained information during conversations with colleagues is another important source.

Since informal verbal communication among scientists is apparently necessary in performing the scientific role, an important question is what influence does position in the social structure of science have on the probability of obtaining such useful information? This is asking about the scientist's opportunities for access to the flow of information and thus has implications in determining his ability to do original research and his ability to compete successfully for scientific recognition. This chapter will therefore focus on two aspects of communication in the HEP community. What are the characteristics of the HEP who use various methods of obtaining information relevant to their research, and how is the informal communication system organized and to what extent is the whole community included (which determines its efficiency)?

Scientific and Structural Effects on Methods of Communication Used by HEP

The HEP were asked to name the most important method they used to obtain information needed to carry out their research. Virtually all the answers were classifiable into either verbal contacts, listening to conference or seminar speakers,[3] or published materials, including preprints and reports. Verbal contacts and published sources were considered equally important.

Libbey and Zaltman (1967:45) report that among 977 theoretical high energy physicists (of many nationalities) journal articles are the most important source of information, with face-to-face discussions ranking second, and "copies of oral presentations (including lecture notes and conference proceedings)" ranking third. This suggested that theorists are different from experimentalists in their information needs and habits of obtaining information.

As with Libbey and Zaltman's sample of theoretical high energy physicists, British theoretical scientists more than experimentalists use published sources of information. Experimentalists rated published sources most important only half as often as theorists did (see table 28). Theorists also indicate more often than experimentalists that listening to conference or seminar speakers is their most important method of obtaining information. There is a simple explanation for this finding. Experimentalists and theorists need different types of information. Experimentalists need to

TABLE 28

MOST IMPORTANT METHOD OF OBTAINING INFORMATION RELEVANT TO RESEARCH, BY TYPE OF HEP

Most Important Method of Obtaining Information Relevant to Research	Type of HEP Experimentalist	Theorist	Total
Verbal	60	21	46
Conference/seminar	6	10	8
Publications	34	69	46
Total	100	100	100
(*Base*)	(128)	(70)	(198)

$X^2 = 27.49$ (2 *df.*, *prob.* <.001)

know who is doing what and what new developments are likely to occur in which area. This type of information travels by word of mouth. Theorists need full information about what has already been worked out in order that they may attempt some refinement or extend in some way other scientists' work. Complex formulas and proofs have more utility when a theorist can look at them and study them for as long as necessary. Experimentalists, on the other hand, can easily remember if X or Y is going to do Z experiment. Because accelerator designs and capabilities are well known among experimentalists, specific experiments (while they obviously differ in some techniques) may be envisioned without too much difficulty. The best example of why experimentalists do not rate published materials as high as theorists is shown by an experimentalist's remark:

> Publications are always too late. By the time a thing is published you ought to have known about it for six months. To be honest, I never read journals for this reason. I assume that anything that is worth my knowing has already come to my attention—big-headed—but that's life.

Information that theorists need to know, on the contrary, may not be public until in preprint or published form.

Personality differences may explain the importance in

methods of communication. Theorists, according to common folklore, are "thinkers," that is, more introverted, while experimentalists are "doers," or more extroverted. It seems plausible that personality factors might affect methods of communication since it is true that some experimentalists and theorists rate methods of communication similarly. There are no data on personality factors with which to test this explanation, but it seems that research requirements are the most important factor in acquiring information.

It might be expected that highly productive HEP would rate one method of obtaining information higher than another method chosen by HEP with low scientific productivity, and the same differences could be expected for recognition; but no differences exist on the methods of obtaining information between levels of productivity and recognition. If social structure within science—scientific productivity and recognition—does not differentiate on communication channels, other social structure variables such as professional age and rank do have an effect on methods of obtaining information.

Older HEP more than younger HEP rate information obtained through verbal contact over that obtained through published sources (see table 29). Older HEP more than younger HEP rate information obtained at conferences and seminars as an important source of information. This is similar to the findings of Garvey and Griffith (1966) that the predominate group in psychology who obtain information through preprints are young doctorate holders and graduate students. A scientist's rank is related to the most important method of communication. Professors (and their laboratory equivalents) more than the other ranks rate verbal sources as the most important method of obtaining information (see table 30). Professors are not more likely than senior lecturers/readers to rate conferences or seminars as the most important source.

Because interrelationships exist between rank and pro-

TABLE 29

MOST IMPORTANT METHOD OF OBTAINING INFORMATION RELEVANT TO RESEARCH, BY PROFESSIONAL AGE

Most Important Method of Obtaining Information Relevant to Research	Professional Age*		
	Younger	Older	Total
Verbal	43	53	46
Conference/seminar	4	14	8
Publications	53	33	46
Total	100	100	100
(Base)	(126)	(72)	(198)

$X^2 = 10.85$ (2 *df.*, *prob.* $< .01$)

*Younger HEP received their Ph.D.s in 1960 or later. Older HEP received their Ph.D.s before 1960.

TABLE 30

MOST IMPORTANT METHOD OF OBTAINING INFORMATION RELEVANT TO RESEARCH, BY RANK*

Most Important Method of Obtaining Information Relevant to Research	Rank			
	Lecturer	Sr. Lect./Reader	Professor	Total
Verbal	44	46	65	46
Conference/seminar	4	19	12	8
Publications	51	35	24	46
Total	99	100	101	100
(Base)	(144)	(37)	(17)	(198)

$X^2 = 13.96$ (4 *df.*, *prob.* $< .02$)

*Laboratory HEP were assigned academic ranks for purposes of analysis. Scientific officers and senior scientific officers were assigned "Lecturer" rank. Professional scientific officers were assigned "Sr. Lecturer/Reader" rank. Senior professional scientific officers were assigned "Professor" rank.

TABLE 31

MOST IMPORTANT METHOD OF OBTAINING INFORMATION RELEVANT TO RESEARCH, BY RANK AND TYPE OF HEP

Most Important Method of Obtaining Information Relevant to Research	Type of HEP: Experimentalist Rank: Lect.	Sr. Lect./Reader	Prof.	Theorist Rank: Lect.	Sr. Lect./Reader/Prof.*	Total
Verbal	61	52	70	14	42	46
Conference/seminar	3	16	10	6	21	8
Publications	36	32	20	80	37	46
Total	100	100	100	100	100	100
(Base)	(93)	(25)	(10)	(51)	(19)	(198)

X^2 (*Experimentalist*) = 6.51 (4 *df.*, *prob.* <.20)
(*Theorist*) = 12.22 (2 *df.*, *prob.* <.01)
(*Table*) = 45.34 (8 *df.*, *prob.* <.001)

*The senior lecturer/reader category was combined because of small N's; while this combination would have been incorrect for experimentalists since a "U-shaped" curve exists across ranks for verbal communication, it is correct for theorists since the increase across ranks for verbal communication is consistent.

fessional age and professional age and type of HEP (experimentalists are older), controls are necessary to determine what the actual relationship is between communication methods and these variables. Controlling for type of HEP, the relationship between rank and most important method of communication is reduced but still present (see table 31).[4] Controlling for type of HEP, professional age is not related to the most important communication channel for experimentalists but is so for theorists (see table 32).

An interaction effect exists between type of HEP (which is role differentiation between theorists and experimentalists) and professional age (which is a structural variable). More experimentalists use verbal sources than theorists; more theorists use published sources than experimentalists, but older theorists use verbal sources significantly more than younger theorists. Regardless of professional

TABLE 32

MOST IMPORTANT METHOD OF OBTAINING INFORMATION RELEVANT TO RESEARCH, BY TYPE OF HEP AND PROFESSIONAL AGE

Most Important Method of Obtaining Information Relevant to Research	Type of HEP: Experimentalist		Theorist		
	Professional Age		Professional Age		
	Younger	Older	Younger	Older	Total
Verbal	60	60	16	35	46
Conference/seminar	4	10	4	25	8
Publications	36	31	80	40	46
Total	100	101	100	100	100
(Base)	(76)	(52)	(50)	(20)	(198)

X^2(*Experimentalist*) = 1.79 (2 *df.*, *prob.* $<.50$)
(*Theorist*) = 12.04 (2 *df.*, *prob.* $<.01$)
(*Younger*) = 25.17 (2 *df.*, *prob.* $<.001$)
(*Older*) = 4.49 (2 *df.*, *prob.* $<.20$)
(*Table*) = 43.36 (6 *df.*, *prob.* $<.001$)

Note: Younger HEP received Ph.D.s in 1960 or later; older HEP received Ph.D.s before 1960.

age, experimentalists use verbal sources in the same proportions. At least two explanations occur to suggest why younger and older theorists differ in their use of verbal sources while younger and older experimentalists do not. Older theorists know more scientists with whom they may discuss research. One way to meet other researchers who may become future communicators is at conferences. The product-moment correlation between professional age and number of conferences attended in the 1966—67 academic year is .11 for experimentalists and .30 for theorists. Attending conferences may not in itself provide communicative opportunities but may result in a second possible explanation, namely, that older theorists are more aware of the key scientists to seek out for helpful discussions and are able to choose those scientists with whom they may discuss research. (Secrecy, or the unwillingness to discuss one's own research, is not related to professional age.) In addition to the low correlation between professional age and number of conferences attended, experimentalists actually spend a significant proportion of their time at the laboratories where they can meet and talk with a large number of different researchers and are more often exposed to other HEP.

Experimentalists and theorists differ as to the most important method of obtaining information relevant to their research. Another question of interest is, if scientists use journals at all, which journals are the most important and why? There are many journals that HEP find relevant, but the most important ones are ranked as follows: (1) *Physical Review Letters;* (2) *Physical Review;* (3) *Physics Letters;* (4) *Nuovo Cimento.*[5] *Physical Review Letters* is a weekly journal that publishes short letters and has a section of abstracts from papers that have been accepted for publication in *Physical Review. Physical Review* is a weekly journal, but the time lag between acceptance and actual publication is much longer than for *Physical Review Letters.* Both of these are publications of the American Physical Society.

Physics Letters, published in the Netherlands, is similar to the American *Physical Review Letters. Nuovo Cimento,* a publication of the Italian Physical Society, also has a "letter" section and publishes in the language of the author so that in any given issue there may be papers in several different languages.

Just as differences between HEP exist in their most important method of communication, the relative importance of journals by HEP provides further evidence to substantiate the interpretation of the information needs of theorists and experimentalists. For the most important journal, 180 HEP chose either *Physical Review Letters* or *Physical Review.* Experimentalists more than theorists chose *Physical Review Letters* as the most important journal. This relationship holds when controlling for professional age. Theorists need the whole results while experimentalists essentially want the news. This interpretation is further substantiated by considering the second most important journal: 169 HEP chose either *Physical Review Letters, Physical Review,* or *Physics Letters.* Here an interesting development occurs. Experimentalists choose *Physics Letters*—an inferior source of news as compared to the American *Physical Review Letters* but still news—while theorists choose the American *Physical Review Letters* (see table 33). This suggests that theorists are looking for quality work, even if the research is not presented as fully as it will be when the complete paper is presented in other journals.[6] This does not refute the different information needs of theorists and experimentalists, but it does show that experimental—as compared to theoretical—work published in non-American journals is probably roughly equivalent to American standards.

Informal Verbal Communication between HEP

Although verbal communication may be the most important method of obtaining information for experimentalists,

TABLE 33

JOURNAL RATED MOST IMPORTANT, BY TYPE OF HEP

Most Important Journal	Type of HEP Experimentalist	Theorist	Total
Physical Review Letters	83	38	68
Physical Review	17	62	32
Total	100	100	100
(Base)	(120)	(60)	(180)
Journal Rated Second in Importance			
Physical Review Letters	16	52	27
Physical Review	15	21	17
Physics Letters	69	27	56
Total	100	100	100
(Base)	(117)	(52)	(169)

X^2 (*Most important journal*) = 37.43 (1 *df.*, *prob.* < .001)
(*Journal rated second in importance*) = 29.26 (2 *df.*, *prob.* < .001)

it is used to some extent by both types of HEP, and the extent of theorists and experimentalists talking with each other is of special interest. (Information on communication behavior was not obtained from two HEP, and seventeen others cannot talk with their counterparts in theoretical or experimental research within their own university since their colleagues are of the same type as themselves. These nineteen HEP will be eliminated from this analysis.)

The reason for considering communication across the experimental/theoretical line stems from the way research is organized. In any institution with one type of experimental research (i.e., bubble chamber or counter), all HEP work on the same team. Collaboration in experimental HEP only occurs between groups or teams; to ask an experimentalist if he communicates with his teammates would not result in meaningful data. While theorists are not formally organized in the same manner, collaboration frequently occurs; staff members usually work on related

TABLE 34

VERBAL COMMUNICATION WITH COUNTERPART (EXPERIMENTALIST/THEORIST) IN OWN INSTITUTION, BY PROFESSIONAL AGE

Communicates with Counterpart	Professional Age: Younger	Older	Total
No	48	15	36
Yes	52	85	64
Total	100	100	100
(Base)	(79)	(46)	(125)

$X^2=13.64$ (1 *df. prob.* <.001)
gamma =.68

Note: Thirty-two percent of each professional age category do not communicate with another HEP inside their own institution: therefore "communicativeness" is controlled in this table.

TABLE 35

VERBAL COMMUNICATION WITH COUNTERPART (EXPERIMENTALIST/THEORIST) IN OWN INSTITUTION, BY TYPE OF HEP

Communicates with Counterpart	Type of HEP: Experimentalist	Theorist	Total
No	23	59	36
Yes	77	41	64
Total	100	100	100
(Base)	(81)	(44)	(125)

$X^2=15.71$ (1 *df., prob.* <.001)

Note: Thirty-seven percent of the experimentalists do not communicate with another HEP inside their own institution. Part of this is because in some institutions there are no other HEP present except their teammates. Twenty percent of the theorists do not communicate with any HEP in their own institution.

problems so there is frequent communication. Moreover, information about communication with others about HEP was desirable—not simply conversation between friends.

Academic rank, professional age, and type of HEP proved important in considering the most important method of communication. A scientist's rank is related to whether he talks with someone other than his own type, but if professional age is held constant, academic rank and verbal communication with counterparts are not related. Both professional age and type of HEP are related to whether a scientist talks to other than his own type. Older men more than younger men talk to their counterparts (see table 34), and experimentalists more than theorists cross the experimental/theoretical line (see table 35). There is an interaction effect between type of HEP and professional age (see table 36). More of the older experimentalists talk to their theoretical colleagues, younger experimentalists and older theorists talk frequently with their counterparts, but few younger theorists talk to their experimental colleagues. Younger experimentalists obtain their most important information while talking, but

TABLE 36

VERBAL COMMUNICATION WITH COUNTERPART (EXPERIMENTALIST/THEORIST) IN OWN INSTITUTION, BY PROFESSIONAL AGE AND TYPE OF HEP

Communicates with Counterpart	Professional Age* Younger Type of HEP		Older Type of HEP		
	Exp.	Theorist	Exp.	Theorist	Total
No	35	67	9	36	36
Yes	65	33	91	64	64
Total	100	100	100	100	100
(Base)	(46)	(33)	(35)	(11)	(125)

$X^2 = 24.93$ (3 *df.*, *prob.* $< .001$)

*Younger HEP received their Ph.D.s in 1960 or later. Older HEP received their Ph.D.s before 1960.

the communication of older theorists is explained by their ability to discuss the broader ranges of research. There is a type of communication gap between theorists and experimentalists, and older theorists are more able than younger theorists to bridge that gap.

There is no difference in the proportion of theorists and experimentalists who communicate outside their own institution; when considering communication outside the institution, the main concern again will be crossing the experimental/theoretical line. In contrast to verbal communication within one's institution, the productivity and recognition variables show significant relationships with communication outside one's institution. Rank and professional age are also related to communication across the line outside one's institution, but type of HEP has no relationship—theorists are just as likely as experimentalists to have verbal contacts with the other type outside their own institution. The interrelationships between these variables require some further clarification.

Rank may be dismissed as having any effect of its own since controlling for professional age eliminates its relevance. The simple effect of experience is present. Older HEP know more people, have established contacts, and are able to name HEP with whom they regularly communicate. This is not saying that they are asked for information or that they are able to give information but only that they communicate with those individuals. Professional age has an independent effect on verbal communication (see table 37).

Productivity and recognition are related to whether a scientist communicates verbally with the other type outside his university. It will be recalled that the productivity and recognition indices are standardized for professional age, so controlling for professional age has no effect on the relationship between either productivity or recognition and communication with counterparts outside one's institution. Productivity and recognition were shown to

TABLE 37

VERBAL COMMUNICATION
WITH COUNTERPART (EXPERIMENTALIST/THEORIST)
OUTSIDE OWN INSTITUTION, BY PROFESSIONAL AGE

Communicates with Counterpart	Professional Age		
	Younger	Older	Total
No	80	48	66
Yes	20	52	34
Total	100	100	100
(Base)	(55)	(44)	(99)

$X^2 = 11.29$ (1 *df.*, *prob.* $< .001$)
gamma $= .63$

Note: Fifty-seven percent of younger HEP do not communicate with HEP outside their own institution; 40 percent of older HEP do not communicate with HEP outside their own institution. Therefore, professional age is related not only to communication itself but also to whether the communication crossed the experimental-theoretical line.

TABLE 38

VERBAL COMMUNICATION WITH
COUNTERPART (EXPERIMENTALIST/THEORIST)
OUTSIDE OWN INSTITUTION, BY PRODUCTIVITY AND RECOGNITION

Communicates with Counterpart	Productivity				
	Low Recognition		High Recognition		
	Low	High	Low	High	Total
No	74	75	71	32	66
Yes	26	25	29	68	34
Total	100	100	100	100	100
(Base)	(46)	(20)	(14)	(19)	(99)

$X^2 = 12.16$ (3 *df.*, *prob.* $< .01$)

be highly related in chapter 4. The result of controlling for productivity and examining the effect of recognition on verbal communication shows that only when there is both high productivity and high recognition is there a substantial difference on verbal communication (see table 38). Scientists with this combination of traits more than any other combination communicate across the experimental/theoretical line outside their institution.

To summarize, verbal communication within one's institution is a function of type of HEP and professional age. Verbal communication beyond the confines of one's institution is a function of professional age and being a highly productive and recognized scientist. All of these variables are structural in the sense that professional age is a synonym for experience and knowledge of the system and its participants. Recognition and productivity are structural in the sense that high productivity is an achievement that sets one apart from his peers while high recognition grants status to the deserving scientist. He almost has a right to approach any fellow researcher and talk with him.

Communication Networks

Cole and Cole (1968) studied the communication system in American physics and concluded that it is efficient apparently because most scientists are aware of other scientists' relevant research. While their study allowed all types of "awareness," ranging from only having seen another scientist's paper to being in a collaborative relationship, the focus here is on regular verbal communication. Similarly, it might be concluded that efficient communication exists within the HEP community if most scientists are in verbal contact with most other scientists.

Contact with other HEP does not have to be at the person-to-person level in order to be effective. If scientist A talks to scientist B who talks to C, then A is, to some extent, in contact with C. The chain of contacts may have several levels of directness. In this study, scientists named others outside their own institution, and these data (after assigning a number to each name) were analyzed by a computer program developed by Coleman (1964) that finds the direct and indirect connections between, in this case, HEP who "talk" and those who are "talked to" by others. (See Crane [1969] for an example of this procedure and for references to other research utilizing it.)

Part of the output of the program provides information on the proportion of potential connections between members of the HEP community who are actually connected (at any level) through verbal communication. To give a brief example, suppose a small group of five people are asked about their intragroup communication. If person 1 talks to person 2 who talks to person 3 (etc.) until a complete circle is reached, the matrix would be thus:

		One Talked To				
		1	2	3	4	5
	1		x			
One	2			x		
Who	3				x	
Talks	4					x
	5	x				

There are twenty possible connections in this group: the number in the group times the number in the group minus one (to eliminate connections with self). At the direct level of connectivity, since each talks to only one other person, there are 5/20 connections. At the level of indirect connections, there are 20/20 since each is connected at some level to all the others. Dividing the number of actual indirect connections by the potential indirect connections gives a measure termed the *connectivity score.*

Information on subgroups of the community may be

analyzed in two dimensions. One dimension is the proportion of potential connections between the groups and the total community that are realized through their own communication. Different subgroups may be compared to determine the extent of their connections to the total group through the persons they name. Another dimension is the proportion of connections based on others' choices, and subgroups may be compared to determine their total connections based on others' having chosen them; for example, some groups may name others, but the choices may not be reciprocal.

Structural and Scientific Bases of the Informal Communication Network

Connectivity scores are used in two ways: the absolute size gives an indication of total connectivity; the "difference" between the absolute sizes of "own choices" and "choices by others" shows the extent to which a group is named more or names more (see table 39). Of all groups, the greatest absolute connectivity score based on "own choices" is the counter/spark chamber group, which has about 4 percent of all potential connections realized. The largest score based on "choices by others" is for the phenomenologists, who have about 5 percent of all possible connections realized. Phenomenologists have the second highest score (after counter/spark chamber) based on "own choices" and also have the largest positive "difference" score. Counter/spark chamber physicists are the only group whose choices connect them with others more than others' choices connect them.

These results substantiate the explanation of why theorists more than experimentalists are highly recognized in the community: theorists provide information in their research that is more useful to the community. These data suggest that theorists also provide more useful information through informal communication channels.

TABLE 39

CONNECTIVITY SCORES* BASED ON REGULAR VERBAL COMMUNICATION FOR EACH TYPE OF HEP

	Experimentalist			Theorist				
Scores Based on:	Counter/ Spark Chamber (80)**	Bubble Chamber (52)	Sub Total (132)	Phen. (18)	Interm. (44)	Abstr. (9)	Sub Total (71)	Sample Total (203)
Choices by others	2	1	2	5	1	0	2	2
Own choice	4	0	2	3	0	0	1	2
Differences	−2	+1	0	+2	+1	0	+1	0

*Connectivity scores are the proportion (expressed here as percentages) of potential connections (at whatever level of directness) which are realized either "Choices by Others" or "Own Choices." The number of potential connections for the total sample is (the sample size) *times* (itself-minus one, which eliminates self-choices). For subgroups, the number of potential connections is (the number in the subgroup) *times* (the number in the total group minus one to eliminate self-choices).
**Figures in parentheses refer to the number of HEP in that category.

Scientists who are both highly productive and have received high recognition from the community are the most likely to communicate across the theoretical/experimental line outside their own department. The connectivity scores show a different structural pattern when considering informal communication with either type of scientist (see table 40). HEP with low productivity and high recognition are connected to more HEP than any other group through their own choices as well as through others' choices. This suggests that communication is organized along structural lines based first on prestige and second on productivity. The second group in the extent of connectivity is high in both productivity and recognition. The third group is comprised of high producers who are low in recognition. This group is apparently not as useful to the community as the first two but more useful than the group fourth in connectivity, those low both in productivity and recognition.

The communication system is organized differentially around the subgroups of types of theorists and types of experimentalists that have been termed scientific bases. It is also organized around the productivity and reward structure, so a look at both bases at the same time is necessary (see table 41). Phenomenologists, in all except the low productivity and recognition category, are connected

TABLE 40

CONNECTIVITY SCORES BASED ON REGULAR VERBAL COMMUNICATION FOR PRODUCTIVITY, BY RECOGNITION

	Productivity				
	Low Recognition		High Recognition		
Scores Based on:	Low	High	Low	High	Total
Choices by others	1	4	1	3	2
Own choice	2	2	2	1	2
Differences	−1	+2	−1	+2	0

TABLE 41

CONNECTIVITY SCORES BASED ON REGULAR VERBAL COMMUNICATION FOR EACH TYPE OF HEP, BY PRODUCTIVITY AND RECOGNITION

	Productivity					
	Low Recognition		High Recognition			
Type of HEP	Low	High	Low	High	Total	(N)
Phenomenologist						
choices by others	1	14	2	5	5	(18)
Own choice	3	4	0	2	3	
Differences	−2	+10	+2	+3	+2	
Intermediate						
choices by others	1	0	0	2	1	(44)
Own choice	1	0	0	0	0	
Differences	0	0	0	+2	+1	
Abstract						
choices by others	0	0	0	0	0	(9)
Own choice	0	0	0	0	0	
Differences	0	0	0	0	0	
Bubble chamber						
choices by others	0	0	2	4	1	(52)
Own choice	0	0	0	0	0	
Differences	0	0	+2	+4	+1	
Counter/spark chamber						
choices by others	1	6	2	4	2	(80)
Own choice	3	4	3	5	4	
Differences	−2	+2	−1	−1	−2	

to more HEP through being the choice of others. Intermediate theorists, in order to be connected more through others' than their own choices, must be high both in productivity and recognition. Abstract theorists are not participants in the communication system. Bubble chamber experimentalists are connected more by others' choices only if they are high producers. Finally, counter/spark chamber experimentalists must have high recognition to be connected more by others' choices.

The explanation for this seems to result from the nature of phenomenological research. Phenomenologists bridge the gap between experimental and theoretical work. They are familiar with both the experimental and theoretical aspects of elementary particles. They know others and are known. They are capable of interpreting the progress of research both to theorists and experimentalists. That phenomenologists' scores are almost as large as counter/spark chamber HEP, with the latter's obvious exposure to other HEP at laboratories, substantiates their function as mediary.

The size of the connectivity scores for counter/spark chamber experimentalists is explained by their relative density around central laboratories. Bubble chamber experimentalists are not required to go to the laboratories nearly as often. At a lab it is relatively easy for scientists to meet colleagues from other institutions, and in addition to other university scientists there are permanent staff members at the labs.

Price (1963) characterized scientists at the forefront of research and in frequent contact with each other as members of "invisible colleges." An invisible college represents a particular form of communication organization. One type of invisible college is in the international community of a research specialty. Another type is possible within a national scientific community, and that is the focus here since data are available only on the British segment of this international specialty.

Since the concept of "invisible colleges" is referred to as an hypothesis the concept is in doubt—at least for some observers. It may be that the formation of invisible colleges occurs in some but not all science specialties, but that is an empirical question. Criteria may be specified for testing the invisible college hypothesis. An invisible college will be deemed present if two conditions are met: (1) if communication choices by the total community lead to a small number of HEP who are named more frequently than the mean times others in the community are named; and (2) if that small group communicates more with its own members than it does with HEP in the larger com-

TABLE 42

NUMBER OF TIMES HEP CHOSE AND WERE SELECTED AS REGULAR COMMUNICATORS OUTSIDE OWN INSTITUTION

Number of Choices HEP Made	Number of Times HEP were Selected									Total	Percent
	0	1	2	3	4	5	6	7	8		
0	104	19	12	1	1	1	1	1	—	140	69
1	25	5	1	—	1	—	—	—	—	32	16
2	6	3	4	—	—	—	—	—	—	13	6
3	3	—	—	1	1	1	—	—	—	6	3
4	2	—	1	—	—	1	—	—	1	5	2
5	2	—	—	—	—	—	1	—	—	3	2
6	1	—	—	—	—	—	2	—	—	3	2
7	—	—	—	—	—	—	—	—	—	0	0
8	—	—	—	—	—	—	—	—	—	0	0
9	—	—	—	1	—	—	—	—	—	1	1
Total	143	27	18	3	3	3	4	1	1	203	101
Percent	70	13	9	1	1	1	2	1	1		99

r = .87; *gamma* = .32

Note: Each cell represents HEP with particular combination of choices and selections.

munity. The first condition assures that the group considered in condition (2) is not a mutual admiration society but rather made up of leaders in research who are in touch with other members of the scientific community. Condition (2) is necessary to ensure that the HEP named by the whole community under condition (1) are in fact a "group" and not simply an aggregate of the more popular people that subgroups have nominated.

To test condition (1) the distribution of choices is examined. The 203 HEP named 138 British HEP as being regular communicators outside their institution (see table 42). These 138 choices are mainly directed to a small group of people; 31 percent of the HEP were selected for all the nominations (30 percent made all choices). The correlation between the number of other HEP named and the number of times a HEP is named is highly correlated (product-moment correlation equals .87). The raw distribution in table 42 shows the extent to which a small group is responsible for naming and being named most of the time. That table provides substantiation for the first condition, that communication choices lead to a relatively small group of HEP.

It could be argued that the concentration of communication is a result of lecturers talking to professors who in turn talk to each other; that is, that communication is for "political" rather than scientific reasons. There is a relationship between the rank of the people talking and the HEP being talked to (see table 43), but the relationship (gamma=.34) is only moderate. Rank of chooser and chosen does have interesting distributions. Lecturers and professors are more likely to choose readers while readers are more likely to choose professors. Although ranks of choosers and chosen are related, it would be erroneous to stress the rank effect for it is the case that lecturers do talk to professors and readers do talk to lecturers (although only once did a professor name a lecturer).

Returning to table 42 to examine in detail those

TABLE 43

RANK OF HEP WHO CHOOSE COMMUNICATORS AND ARE SELECTED FOR COMMUNICATION OUTSIDE OWN INSTITUTION

Rank of HEP Selected for Communication	Rank of HEP Choosing HEP Communicator Outside Own Institution: Lecturer	Sr. Lect./Reader	Professor	Total
Lecturer	33	28	4	27
Sr. Lect./Reader	46	28	57	42
Professor	21	44	39	31
Total	100	100	100	100
(Base)	(72)	(43)	(23)	(138)

X^2=14.20 (4 *df.*, *prob.* <.01)
gamma =.34

Note: It should be emphasized that this table relates to choices or selections and not to individuals; 138 choices are involved but only 99 HEP.

individuals falling in cells whose value on either axis is three or greater shows the following: these 24 HEP made 59 percent of the choices (81 out of 138); and 16 of the HEP received 56 percent (77 out of 138) of the choices. The question is—in order to test condition (2)— do these HEP choices lead more to themselves or to HEP out of the group? Examination of their choosing and choice matrix shows that 58 percent of this group's choices lead to other group members (indirectly, every member is in contact with every other member), and of the 77 choices that led to this group, 62 percent originated from this group. Taking into consideration the large number of others who could be selected, there is ample evidence to suggest the presence of an invisible college.[7]

The question of efficiency in the communication system includes an analysis of connectivity scores for research groups (bubble chamber, spark chamber/counter, and theory groups) and institutions. Group and institutional analysis has the same outcome as individual analysis if connections on both communication and being a colleague are included as a communication tie.[8] Rather than consider individuals, therefore, groups (and institutions) will be used as units of analysis. This has the effect of treating all groups as though they are equal in size when, of course, they are not. Two groups with ten members each have one hundred possible ways of being connected to each other whereas two groups with five and ten members each have only fifty possible ways. Although the question of whether there are communicative advantages in being a member of a larger group or institution is relevant, the interest here is in whether groups are in fact connected by communication ties.

The bubble chamber group subset is both the smallest in number and the least connected (see table 44). Counter/spark chamber groups are the most connected. Theory groups—the largest subset in number of groups—are between bubble chamber and counter/spark chamber groups in connectivity. There is a simple explanation for the different connectivity scores for each group. Counter/spark chamber groups conduct their experiments at

TABLE 44

CONNECTIONS* BETWEEN SUBGROUPS OF THE HEP COMMUNITY

Group	Number of Groups	A** Potential Connections	B*** Actual Connections	C# Connectivity Scores
Bubble chamber	11	110	9	.0818
Counter/spark chamber##	16	240	184	.7666
Theoretical	19	342	85	.2485
All institutions	23	506	308	.6087

Note: The total number of different groups is 46; however, several institutions have more than one group so the "All Institutions" line in this table is not the summation of the other lines.

*Connections are based on anyone at a given institution having regular communication with anyone at another institution.

**Potential connections equal the number of groups times the number of groups (minus one).

***Obtained from the connectivity computer program output.

#Connectivity scores are computed from B/A.

##The various groups at the laboratories are considered as one kind of group at each laboratory.

the two national laboratories; they have many opportunities to meet and in fact do so frequently. The laboratory sponsors seminars, discussions, and provides a central dining room for meals as well as a hostel for scientists who stay overnight at the laboratory. At any time there may be guests from a variety of institutions.

In contrast to counter/spark chamber groups, most bubble chamber film comes from CERN so that a national laboratory is not the focus of bubble chamber physicists' experiments. Rather, their own scanning, measurement, and computer rooms where film is analyzed are their "laboratories."

Theory groups' connectivity is between the experimental groups. While they do not have a laboratory, they do have regular seminars in various institutions to which visitors come and go. These are not as conducive to wide exposure as the laboratories but more so than the situation that exists for bubble chamber groups. There has been some informal discussion about establishing a theoretical research center. If it is established, one could expect the same result in the communication system for the theorists as now exists for the counter/spark chamber groups.

The utility of laboratories in fostering communication is not surprising, and is further substantiated by the data (in table 41) where phenomenologists are shown to be linked by others' choices to more scientists than any other theoretical (or experimental) group. Phenomenologists are frequent visitors to laboratories because it is from experimentalists that they obtain their raw data used in analyzing their phenomenological models of elementary particle behavior. To be successful at getting new data requires ability to communicate with experimentalists. It might be advantageous for the progress of research if all theorists had to "court" experimentalists.

An answer to the question of efficiency in the whole communication system can now be suggested. Unless degrees of magnitude are measured against some absolute standard, interpretation should result from a comparison between two systems. In the absence of a comparative system, one must conclude that the communication system in British HEP is highly efficient. The connectivity score (see table 44) indicates that 61 percent of all possible ties are realized, which is a reasonably high figure. Twenty institutions—out of twenty-three—are each connected with at least sixteen other different institutions through their own choices. Only three institutions are not connected to any other through their own choices, but two of these three are connected with twenty-one other institutions through others' choices. This means that, considering connections as a result of choosing or being chosen, only one institution is totally isolated (and it has only three HEP staff members). While less than 2 percent of the individual scientists appear to live in a communication vacuum, 98 percent are connected at some level, which is sufficient to conclude that the communication system in the high energy physics community is efficient.

One major conclusion from these findings is that the social structure of the community is reinforced by the communication system. Highly recognized HEP serve as fountains of information from which other HEP drink. Their status comes from the community, and the community in turn gets the benefit of both their research and their guidance. The community's ability to draw on the experience and knowledge of the best scientists suggests that the system is not only efficient as a system but is also efficient for the progress of science. As with the universalism in the reward system, there are no important problems for scientists' originality and ability to compete revealed in these data. That does not mean that particular scientists in certain instances may not feel they are at a disadvantage in being able to compete, but on the basis of community-wide involvement the informal communication system operates with few restrictions. The apparent restrictions associated with those scientists high in recognition and research productivity having more contact with other scientists are actually positive for scientific progress. Furthermore, these possibilities for access are based on merit rather than on some personal attributes. The informal communication system as organized, in addition to serving well most individuals, enables the whole HEP community to compete with other national research communities.

Summary and Conclusions 8

This study has examined several aspects of originality and competition in the high energy physics community in Britain, ranging from the prestige of research groups in university departments and the reward system to the prevalence and severity of competition and the communication system. Most of these areas have been the topic of separate studies in the sociology of science, but they are related to the theme of originality and competition.

Data from American studies were used for comparisons, and even if they are essentially the only data available, the organizational aspects of the research environment and the educational system are different in Britain and the United States and these differences affect the way science operates. They were described in detail to set the stage for explaining the major findings as follows:

1. Scientific productivity is essentially a function of professional age.
2. Recognition is, first, a function of scientific productivity and, second, a function of the type of HEP. Theorists are recognized more than experimentalists for their research.
3. Competition in HEP is prevalent and is more severe for theorists than experimentalists; but secrecy is not positively related to the probability of being anticipated.

4a. Channels of communication are used differentially by experimentalists and theorists. Experimentalists use verbal sources and theorists use publications. These differences are based on special requirements of each type of research coupled with possibilities for verbal communication based on location in the social structure.

4b. The communication system is organized around the members of the community in an "invisible college" with the flow of communications reinforcing the social structure and vice versa.

5a. The division of labor in the HEP community consistently accounts for large differences in the dependent variables whereas social and educational background and institutional characteristics explain little (if any) of the variation.

5b. The HEP community is divided into theoretical and experimental roles that are further divided into types of experimentalists and theorists. With a highly specialized division of labor, it is significant that the phenomenologists who are familiar with both theory and experiment are the center of the communication system, receiving absolutely the most recognition and also receiving the most recognition for their level of research productivity.

Scientific Productivity

Students in Britain are almost entirely selected for future educational opportunities on the basis of achievement at about age eleven. If selected, they go through the rigorous secondary-school curricula and are admitted to one of the relatively few universities. At the university, it is hoped that they will exhibit intellectual and personal qualities that on the completion of their first degree allow them to remain and do research. Although a large number of students could probably be successful at research, not all are asked or allowed to work for a Ph.D. The few who take Ph.D.s are, therefore, highly selected. It should be remembered that students specialize at a very early age; by the time they are in the sixth form (concentrated secondary school) they are already specializing in a particular science or science and mathematics. The student knows what degree subject he will read for when he comes to a university because of the early commitment, and he finds a competent staff at most if not all of the universities. The undergraduate degree itself is very specialized.

If being located at Oxbridge has traditionally been desirable, not being located there is no indication of mediocre talent or abilities—the available supply of talent is sufficient to stock the Oxbridge staffs as well as the other universities.[1] Consequently, the combination of carefully selected recruits and equally talented teachers appearing in most universities results in a uniformity not found in the total distribution of institutions in the United States. Halsey and Trow (1971:321) write that ". . . an American observer may note that despite the marked differences in the history and function of, say, Manchester and Reading, their membership in a national university system makes them more alike than comparable institutions in the United States." As a consequence of the combination of able students and teachers, the type of institution attended for the Ph.D. (or undergraduate degree) and the prestige of the department are not related to later productivity. When the level of intelligence, motivation, and opportunities for research is so similar, professional age and productivity are closely related. Of course there were exceptions noted—some scientists are highly productive and some have relatively few publications—but, on the average, professional age is the best predictor of scientific productivity. The differences observed between types of universities resulted from the different concentration of specialties in the various universities and the fact that one college in the London University category has a group that formerly was active in a type of nonaccelerator nuclear research. One characteristic of that research was the possibility of producing many short papers. Because of the few cases involved, it was not possible to control for both specialty (five types) and type of institution. The main difference in productivity occurred among those scientists whose research specialties promote differential rates of publication because of the time factor involved in completing a particular type of research. Counter/spark chamber physicists require about two or three years for each experiment and then may publish four or five papers. In the meantime, an abstract theorist is able to publish much more frequently; but even these differences should not be overemphasized since the mean number of publications, controlling for professional age, was not statistically significant between the types of research specialties—the differences were only consistent.

Different opportunities for research are not present to a significant extent in different universities—if research is to be conducted it will be funded as it is in other institutions, although there are categories of levels of funding. The differential funding of bubble chamber groups and the relationship with prestige of departments are explained by the fact that prestige ratings were apparently based on the size of the various groups and by the fact that a particular combination of economic factors and governmental policy negates the possibilities of greatly differential opportunities for productivity. The strong correlation between size of group and prestige ranking coupled with the almost negligible effects of prestige ranking suggests that the size of the group was the main factor.

Economic factors dictate that equivalent funds are not necessary since different sized groups require different levels of funding. For example, three bubble chamber physicists cannot utilize certain equipment to full advantage; but nine can utilize more than three times that amount. As the number of HEP increases, a research effort different in kind—not in degree—occurs. The Science Research Council requires small groups to be affiliated with large groups as "extensions." The small groups may work autonomously or may collaborate, but the major expensive facilities they need must be used at the larger groups' laboratory. Therefore, part of the expense of a small group is hidden in the costs of a larger group. Moreover, in large bubble chamber groups there may be scientists A, B, C, D, and E. Any combination of these may work on different projects (that is, analysis of different types of film) so that A, B, and C, or B, C, and D, are working together at any one time. That resembles the smaller groups that are equivalent to one subset of the larger group, and the smaller group therefore can produce research as quickly as the subset of a larger group.

In summary, productivity and professional age are closely related because of the standardized Ph.D. product and the similar conditions found in various universities, including the quality of the staff and the equivalency of the funding, since the Science Research Council provides centralized planning and funding.

Scientific Recognition

The reward system operates in a universalistic fashion. Recognition is highly related to productivity and to the type of HEP, with theorists receiving more recognition than experimentalists. This effect will be discussed later. Affiliation with a particular type of university or a department in a particular prestige category has no effect on the opportunities to receive recognition for having contributed to the scientific community. The question is, then, why is this different from the United States? The answer to that question involves the methodological problem of small variations on the prestige variable and the characteristics of the system of higher education. Methodologically, the British university system should be compared to the top twenty universities in the United States rather than to the total distribution of the top one hundred universities. For example, in the United States the variation from the top to the bottom of the top twenty universities is insignificant compared to the differences in the top twenty universities and numbers 81 through 100 on a prestige ranking (see Cartter, 1966). When the variation in one of two variables is greatly reduced, its predictive power is consequently reduced.

Within the British university system there is little personal competition for academic positions and even less institutional competition for staff.[2] (This is much different from competition for recognition in which all scientists must engage to some extent.) The universities are able to acquire staff—indeed must acquire staff—without offering fabulous salaries or other considerations that are characterized by faculty mobility in the United States. One department may not organize a new research group without requests both to the University Grants Committee for staff salaries and to the Science Research Council for research funds. In the United States, where there is both a decentralized system of institutional funding and constituencies and a national decentralization of research funding between several agencies, universities and scientists are in stiff competition, but this is not the case in Britain. The only competition arises from the initial decision to allocate resources since departments do not use funds to start a research group and then become involved in competition with other departments in obtaining government funds. The system is planned—it is not allowed to generate its own steam in a competitive atmosphere.

Studies in the United States show that funds are disproportionately distributed to the more prestigious universities—and less prestigious universities complain about it. The British complaints reported from two staff members (and there were others) resulted from the manner of distributing the total funds and the amount they received relative to their expectations, not from the absence of funds. Indeed, one of the complaints reported came from a scientist in the best financed group in the country, and his opinion was that the smaller groups should get less. He said that the officials "wanted to promote the less effective departments to get going." This contrasts with the situation in the United States where scientists at high prestige universities are not likely to believe that lesser institutions should receive fewer funds but that the solution lies in the general availability of more funds so everyone can obtain more.

Competition ensures a type of evolutionary process whereby the most efficient and most adaptable group(s) is able to show the greatest achievement. Ben-David (1968) suggests that because of the decentralized system American institutions are able to innovate and transform outmoded organizations into more efficient production units. This implies that if British science were decentralized and if British universities were able to compete not only with each other but with international groups as well, there would be a higher achievement by some groups (and, of course, some of the smaller groups would soon disappear). The question is not whether competition has positive or negative effects for either a system or a group; the question is whether the lack of competition leads to an equalizing effect in the opportunities for receiving recognition according to merit. The data suggest that all groups are equally capable, and affiliation with any particular group neither stigmatizes nor hallows its members.

In addition to the lack of institutional competition, personal competition in universities is not severe in Britain. Staff members are given tenure after two or three years on a probationary appointment and are then free theoretically to do only a necessary minimum of teaching and no research at all. Research is the necessary condition for advancement, but unless other conditions are right (available positions, willingness to move, etc.) research is not a sufficient condition for advancement. There is, after all, a finite number of positions, and the salary differences are so small that movement between universities to rise in the academic hierarchy—were there positions available—is simply not worth the cost and effort involved. For over a fourth of the HEP, the present affiliation is the same university where the Ph.D. was taken, and that fraction increases to one-third if laboratory scientists are excluded. There may be a few of the so-called academic entrepreneurs, but engaging in such maneuvering does not lead to the same advantages possible in the United States where research empires are alternately built up and scientists move from one university to another at a higher rank and salary, repeating the process until the desirable income level and environment are achieved. This does not mean that mobility is nonexistent, but rather that the noncompetitive aspects of the British system are such that the fruits of geographical mobility are not great. Without intense interpersonal competition with prizes to be won, most scientists simply get on with their research without spending a significant part of their time and energy planning their next strategic move.

Finally, the number of scientists involved in the community is small, and virtually everyone either knows most others, has heard of them, or knows someone who does know everyone. In a closely related group it is easier to monitor achievements (and failures) so that the rules of the game are easier to enforce. The deserving scientist can be rewarded, and in the unlikely event of deviancy the culprit can be sanctioned. A scientist never knows when trying to move to a new position which members of the community—in addition to the references personally submitted—will be asked to evaluate him. And unless migration is desirable, escape from the system is impossible. Consequently, the universalistic distribution of rewards—not only desirable within the norms of science—is required for the continued, smooth operation of the community.[3]

Originality and Competition for Recognition in the Scientific Community

Slightly more HEP than a sample of American physicists have been anticipated by fellow scientists at home or abroad. One explanation for this is that the specialty of high energy physics is more competitive than physics in general. Another explanation is the possibility that British HEP are less capable of international competition because of the organizational problems involved in collaboration. The data do not provide direct evidence for either of these explanations, although the first explanation appears the more appealing.

The major divergence with previous studies involves the question of secrecy and the probability of anticipation. Whereas Hagstrom (1967) reported a strong relationship between concern about anticipation and secrecy, the probability of being anticipated did not affect whether HEP would be secretive about their research. One explanation for these results is that if the probability is high that a British HEP will be anticipated (and even if he is not able to publish as a result) the lack of personal competition for positions and recognition would result in a different outcome for him compared to his American colleagues. The personal feelings he would have as a scientist, nevertheless, would be of comparable severity.

A second explanation raises a methodological issue on the differences between "concern about anticipation" and "probability of being anticipated." Both variables are related to the number of times scientists have been anticipated, but it is possible that each variable is measuring something different. Since the data show effects different from Hagstrom, it is not possible to determine whether the British are more secretive than Americans *as a result of competition;* but 60 percent of the HEP reported secretive behavior, and so it is not being suggested that HEP are not secretive. Rather, what is suggested is that secretiveness in science has at least two components: that resulting from the competitiveness of science and that resulting from reticence, exemplified by the data on the various stages of research at which scientists would not want to discuss their research. Hagstrom's data show that 32 percent of his sample were secretive, although they were not at all concerned about anticipation, which provides additional support for the hypothesis that secretiveness results from reasons other than currently perceived competition.

Communication

Theorists and experimentalists differ in their communicative behavior. For sources of information theorists utilize publications while experimentalists utilize verbal contacts. Professional age is also important in that more older theorists than younger theorists rate verbal sources as the most important source of information. These findings result from the different information needs of theorists and experimentalists.

The social structure and the division of labor within the scientific community are related to the communication system. Theorists obtain more recognition than experimentalists from the scientific community partly because theorists are the main source of information for the system.

Verbal communication is organized around an "invisible college" whereby a few scientists initiate and receive the major share of communication. The receivers are among those who have achieved the highest recognition, and they themselves initiate communication to scientists with similar characteristics more frequently than they do to less prestigious scientists. Since theoretical work—especially the phenomenological type—is useful to experimentalists (even more than other experimental work) they both rely on theorists for guidance and, with the rest of the community, reward theorists for their useful contributions.

The HEP community is a national community of an international scientific community. Ben-David (1969) suggests that one nation usually serves as the center for scientific research. Since World War II the United States has been the acknowledged center for research in physics. Specifically, Ben-David (1969:13) writes about the conditions in the countries that are secondary centers of research:

> The strategy seems to be somewhat like this. By refraining from competition for the position of leadership, but maintaining at the same time continuous exchange of communication and personnel with the center, a country can (1) avoid risk-taking in science, and (2) concentrate its resources on fields and projects where its chances of success are optimal. This has been, I believe, the policy of the British scientific "Establishment" for the last hundred years, and from the point of view of contributions to basic science, it has been a singularly successful policy.

One question is to what extent the British HEP community maintains contact with the American community. Unfortunately, the relevant questions to provide that data were not asked. The data from regular communication with scientists outside one's institution are not the appropriate data to analyze this contact because, although "regular communicators" includes regular written communication, "regular" communication was interpreted to the HEP as meaning the scientists they communicated with the most (by whatever method). Nevertheless, a few young HEP indicated former fellow students or recent colleagues in the United States. One eminent theorist named no one in Britain but named several people abroad. (No one in Britain named him either.) He is probably a member of the international community, but he provides linkage only through formal guest lectures rather than through informal mechanisms.

Data that do indicate links to the United States include the fact that 25 percent (48 of 195) reported that American HEP are the main influence on their current research and 40 percent (82 of 203) have held research or teaching positions in the United States. Additionally, at any one time several British HEP are abroad either on sabbaticals, visiting professorships, or postdoctoral fellowships.

These scientists function as important links, as one experimentalist noted:

I know my competitors—they're very few. Both groups are American. There is X and Y at ———. Of course we've got our own spy in X's camp. Z, who used to work with us, is now out there. He's keen eventually to come back and work with us again, and he feeds us information in the meantime.

The British are not isolated, but the HEP have to work in close cooperation to maintain a community with viable secondary leadership in the international community. This position is now threatened by increased research costs and the fact that future research will require total financing by international organizations whereas only part of the present effort is thus financed through the CERN international organization. The effect on a national community of moving the focus of research from national labs to international labs will be an important research problem for the sociology of science, and science policy, in the near future.

The Division of Labor

The effects of the division of labor between theoretical and experimental roles are not only powerful but also consistent in every aspect of the social system of this scientific community. The division of labor results from the very specialized tasks that each group of the community performs. Abstract theorists produce axiomatic theories that are not beneficial at present in explaining the behavior of elementary particles. Intermediate theorists produce theories that have more relevance for physical reality but are not yet very useful. Phenomenological theorists build and test models that are extremely useful to experimentalists. Bubble chamber experimentalists analyze photographic film to reconstruct elementary particle interactions. Counter/spark chamber experimentalists monitor their detection devices as the particles are accelerated and collide with targets, providing the secondary particles and events of interest, and also analyze film from visual spark chambers. These five groups comprise a continuum of proximity to physical reality, with the abstract theorists being at the opposite end from physical reality while the counter/spark chamber experimentalists not only record actual events as they are happening but are as near as safety measures allow while the experiment is taking place. Since each of these five tasks is so specialized, it is conceivable that these groups would not need to be in frequent communication with each other because of scientific necessity.

Very few experimentalists can judge the quality of an abstract theory, and few abstract theorists are able to judge the validity of experimental results. Indeed, an abstract theorist would not ordinarily be able to suggest a crucial experiment—and if he were able, the experimentalist might not understand what he was talking about. In such a highly fragmented community one wonders what ties this community together? The data consistently indicate that phenomenological theorists do just that. Phenomenologists receive more recognition than any other group, both absolutely and relatively, but this does not provide the most direct evidence that phenomenologists are the cohesive force in the HEP community. The communication networks are the major support. Phenomenologists are a relatively small subgroup (the smallest after the abstract theorists), but they are linked to more members of the scientific community than any other subgroup because members of the community name them most frequently as scientists with whom they regularly communicate. Even based on their own choices of regular communicators, phenomenologists are connected to other HEP in a proportion almost the magnitude of the counter/spark chamber experimentalists who have large numbers in or near the two national laboratories. Phenomenologists do not have this natural meeting place, but they are successful nevertheless in reaching almost as large a portion of the community through their own communication choices.

There are several implications to be gained from the effects of the division of labor in this scientific community. In a highly specialized community, it is not the most specialized group that has the highest status; in fact, it is just the opposite. Phenomenologists are able to translate experimental findings to other theorists and to provide the same service for experimentalists. Phenomenologists are specialists, but they are specialists at being generalists. Their status is derived from the role of being the "middle man," and from the usefulness of this role for the scientific community.

A second implication of the division of labor involves the extent to which this development will occur in other sciences. As other sciences become so specialized that subgroups within a specialty (not just within a discipline) are actually unable to conduct meaningful communication with each other except through a specialist who is able to grasp the whole subject area, there will be scientists who are anxious to play this middle role because of the status that accompanies it. Ironically, this suggests that instead of science becoming more specialized, forces will operate to pull more scientists into this middle position. Taken to the extreme, this would result in a despecialization of science. One reason for this not happening is the fact that the middle position is an extremely difficult role to play (one is tempted to describe the role as that of a "double agent") because of the breadth of knowledge necessary.

A third implication of the effects of the division of labor is related to the increased specialization in all professions. Projecting the results from a scientific community that is presently very high on a specialization scale, it is possible to predict that when roles become segmented one role will emerge whose function is to bind the separate components into a cohesive group and that the occupants of this middle role will enjoy a status comparable to its utility for the community.

This study has shown that the social organization of scientific research in Britain is different from that in the United States, and these differences affect the internal operation of the community of science itself. The centralization of science policy including centralized funding and research planning is sharply contrasted to what has been termed a "pathological" decentralization in the United States. That science in Britain is able to function well within the normative structure of science has been shown. Britain cannot afford, in resources or scientific abilities, the wastes that result from the much more competitive situation found in the United States. And the question is being raised frequently in the United States whether it can much longer afford the type of science organization it has had—and fared so well under—for the last quarter of a century. If centralization must eventually occur in the United States, and some degree seems inevitable,[4] then the British case is an example of the possibilities of having a scientific community without a rigid hierarchy of a few institutions. That consequence can be avoided, though, only if scientists themselves are adequately represented in the planning and administration of the science policy. To have it imposed by administrative directives would be disastrous. The American and British conditions are much different in both size and historical traditions, so certainly a blueprint from Britain would not be appropriate for American science. But in considering the social implications for the scientific community, those aspects of the British experience, which serve to enhance the internal operation of science through adhering to the norms of science, should be given careful consideration.

It seems improbable that science could ever be so organized that scientists would not have to compete with each other for recognition from the community of scholars engaged in research. Certainly competition would be diminished if agreements were reached, dividing up research into separate areas, and each scientist or group worked only on a specific area. That would involve international agreements, and as such, unless the funding came from a central location with the necessary sanctions available to be applied to those tempted to encroach upon an area assigned to other scientists, it seems unlikely that a gentlemen's agreement would suffice.

On a national scale, research in elementary particles already has come to be organized around such understandings (although there are no written contracts), based on the magnitude of funding as well as the different research possibilities and limitations of various experimental accelerators. Even so, international competition between the United States, western Europe (which involves understandings between different groups comprised of different nationalities at CERN), and the Soviet Union is still energetic. Although a global accelerator is more probable than any other type of cooperative research instrument —except perhaps some type dealing with space—it has not yet been planned. Much further in the future is this type of concentration of research funding and policy for all other scientific fields. In the foreseeable future, then, scientists face the problem of doing original work in competition with others. As long as this happens, the problems of the distribution of rewards and the secretiveness to protect advantages will remain.

When there is more competition among researchers in the social sciences, one can look forward both to the progress that it will herald as well as the excitement it will engender.

Appendix

Prestige, Productivity, and Recognition Indexes

Prestige Index

There are twenty institutions in this study excluding the laboratories, but only eighteen institutions were listed on the rating sheet. One of the eighteen was listed in error, because it has a cosmic ray physics research group that is not included in the study. In the original source of names of the researchers, *Scientific Research in British Colleges and Universities,* scientists describe their own research specialties and these cosmic ray specialists correctly described their research as the study of elementary particles, but the reference did not include the fact that they used cosmic rays. When I learned that this group was listed in error, I left the name on the list as a test. If any HEP had checked other than "insufficient information" for that institution, his ratings would have been excluded. All raters either checked "insufficient information" or questioned the presence of a research group at that institution, only at which time were the scientists informed of the original error and told why the institution was listed. As a result only seventeen of the relevant twenty institutions were rated.

Three institutions were not listed. Some groups that are not in physics departments have been cross-referenced in the physics section of the publication used, but two groups were not. Since some were cross-referenced it was assumed all were and thus they were missed. The third institution not listed had HEP on the staff for the first time in 1967–68 and could not have been listed in the source, which is published at the end of the academic year. Scientists at the institutions omitted from the list, of course, were visited and interviewed.

Blank spaces were provided on the sheet where the institutions were listed with the request that "other" relevant groups be added by the rater. Only two HEP suggested an additional group. Therefore, in the categorization of high prestige, middle prestige, and low prestige, the three groups not printed on the form were arbitrarily classed "low" prestige. That classification seemed sufficiently warranted. A raw score of 2.00 for the mean ranking was assigned to HEP at these three institutions for correlational analysis. This assignment is arbitrary, of course, but not unreasonable in light of the lowest rated institution having a mean rank of 2.06.

Ratings were obtained from 190 of the 203 HEP. Of the 13 nonraters, three were not requested to rate the groups because of time considerations; six had recently returned from a long period abroad and felt they were not qualified raters; and four refused to rate anyone. These last four willingly answered other questions but said they did not want to rate other scientists. All laboratory HEP participated in the rating, but the laboratories were not rated.

Because some groups are small and because the HEP were assured that identities would not be revealed, the individual ratings by name of university will not be presented. If the ratings had been for the complete physics departments rather than research groups the ethical problem would not be relevant. The prestige groups were assigned by choosing approximately one-third of the scientists in each category, although the number of institutions represented in a category was inversely related to prestige.

Some groups are perceived to be more prestigious than others, and there was considerable agreement as to which groups should be rated high. Raters indicated "insufficient information" more when rating the lower ranking groups than when rating the high-ranking groups. In fact, the rank order correlation between prestige rank and number of "insufficient information" responses is −.96.

Productivity Index

To provide comparable data with previous studies, especially Crane (1965) and Hargens and Hagstrom (1967), a productivity index was computed. The correlation between professional age and number of cumulative publications is .65. Therefore, productivity was standardized by taking each professional age group (beginning with HEP who are in the first year after their doctorate, second year, etc.) and assigning one-third of the group with the largest number of publications to the "high" productivity category and the other two-thirds to the "low" productivity category. This procedure is slightly different from Crane (1965). Her criteria for high productivity involved a minimum of weighted publications (in which books were given more weight than articles) for each professional age. In this study, high producers are so vis-à-vis other HEP in the sample, not as measured against a specified standard. Thus, whereas her proportions of high producers varied by professional age groups (from 23 to 57 percent) the proportion for each professional age in the present study is one-third.

There were two exceptions in the procedure of assigning "high" and "low" productivity: (1) No HEP was assigned "high" productivity if the number of his publications was smaller than a HEP assigned "low" publications in an earlier professional age. For example, suppose in professional age group 5 there were six HEP, with two (one-third of six) classified "high" and four classified "low." If the top two had ten and seven publications respectively, but in the lower age group 4 a HEP with seven publications was assigned "low" productivity because one-third of that age group had eight or more publications, then the five-year-old with seven publications would be categorized "low." Then in age group 6 an additional HEP would be categorized "high" to maintain the one-third fraction, provided that he would have been categorized "high" in any previous age group. (2) When in the older professional age categories there were few HEP, two or more successive age groups were combined, but following the same procedure as just outlined.

Experimentalists publish joint papers almost exclusively, some ranging up to fifty coauthors. Theorists publish jointly much less. Therefore, the productivity index was computed separately for theorists and experimentalists to take cognizance of this difference. This procedure was to follow the suggestion of Price and Beaver (1966), that is, assigning a scientist a part-score of 1/n papers (where n = the number of authors) for each publication. That procedure would have been difficult since it requires knowledge of how many coauthors each HEP has on every different paper. Had experimentalists been assigned a score based on the total part-score for each paper, it is likely that many would have had a publication total of less than two or three paper equivalents. If it is assumed, however, that each HEP has the same number of coauthors on the average, the same relationship vis à vis other HEP would obtain. That assumption is not true for theorists and experimentalists together, while it is very likely true for each group separately. Thus, their suggestion was implicitly followed when each group was treated separately.

As a check, both experimentalists and theorists were combined, disregarding any differences in how many coauthors might be involved, and coded again on the basis of cumulative totals for each professional age group. Only nine scientists, four experimentalists and five theorists, had their categories changed, so the procedure actually used was only slightly better than simply grouping everyone together.

Recognition Index

Information was obtained for seven items, which form the basis of the recognition measure: (1) Number of papers refereed (entire career); (2) Number of lectures at other institutions in Britain during the 1966–67 academic year; (3) Number of lectures at institutions outside Britain during the 1966–67 academic year; (4) Number of invited papers during the last two academic years; (5) Membership on a national policy or funding committee (entire career); (6) Editorship or editorial board member (entire career); (7) Membership in honorary societies.

Three kinds of recognition scores were constructed from these data. The first was the number of *different* relevant items, the total score possible being 0 to 7. The results are distributed as follows:

Number of Different Activities	Number of Scientists
0	96
1	45
2	32
3	19
4	9
5	1
6	0
7	1
Total	203

A second, more complicated recognition score was constructed. For each item, the number of scientists involved was divided into the number of times that item was reported. For example, 27 scientists had given a total of 56 lectures outside the country in the previous year. The average number of lectures for each HEP involved was 2.07. For each activity the distribution is as follows:

	Number of Scientists Involved	Number of Activities Reported by Relevant Scientists	Mean for HEP Involved	Rank in Importance
Fellowship, Royal Society	5	5	1.00	1
Editor/Editorial Board	6	8	1.33	2
Committee Membership	21	29	1.38	3
Invited Paper	22	35	1.59	4
Lecture Outside	27	56	2.07	5
Lecture in Britain	65	170	2.62	6
Refereed Papers	68	671	9.87	7

The Royal Society is the main honorary society in Britain for scientists, and it has limited invited membership. The last item, the number of papers refereed, was subject to some variation. When a scientist indicated a range in the number of papers refereed, the lower value of the range was used.

It seemed logical that if many scientists were involved in an activity, the net recognition to each scientist would be less than the recognition from an activity in which only a few scientists were involved. For each item, a scientist was assigned a partial score based on his proportional participation. For example, suppose he (1) was a Fellow of the Royal Society, (2) had refereed twenty-five papers, and (3) had given seven invited papers. His proportional participation score would be:

.200—fellowship in Royal Society 1/5 (since there are five *fellowships* involved).

.037—refereed papers 25/671 (since there were 671 papers *refereed*).

.200—invited papers 7/35 (since there were 35 *invited* papers).

.437—total proportionate score

The third index involved taking the items as ranked above and assigning a number to that activity based on an inverse ranking so that fellowship in the Royal Society had a score of seven, whereas refereeing papers had a score of one. Any scientist, therefore, could have a total score ranging from 0 to 28. Taking the same example as above, a recognition score would be derived as follows:

7—fellowship in Royal Society
1—refereed papers
4—invited paper
—
12—total score

The zero order product-moment correlations between the three types of scores are as follows:

	1	2	3
1. raw number of items	—	—	—
2. total of proportionate items	.75	—	—
3. total of weighted items	.92	.89	—

The third recognition score was used because it is more highly correlated with the other two. These scores are used in regression and correlational analyses.

A recognition index was also computed for comparing previous studies. For each professional age group, the HEP with the top third scores were assigned "high" recognition and the remaining two-thirds were assigned "low" recognition. The same two exceptions outlined for the productivity index above were followed so that no HEP was categorized "high" in any age group if the recognition score would have caused him to be categorized "low" in a younger age group.

Notes

1 Introduction

1. This does not mean there are no degrees of success in a scientific career. Rather, it is a picture of the problem that is most vivid to scientists working in the areas most susceptible to anticipation in discovery.

2. See Berelson (1960), Caplow and McGee (1958), Crane (1965), Cole and Cole (1967; 1968), Hargens and Hagstrom (1967), and Wilson (1949).

Another problem science faces because of these norms and the norms of the larger society in which it operates is the effect of society's stratification system on recruitment and mobility of scientists. Harmon (1965:33–45) shows that, consistently since 1935 in the United States, scientists' fathers have had occupational and educational achievements higher than the population in general. Assuming that potential scientific abilities are distributed among all social classes, scientific potential is lost in not recruiting equally from all social categories. This is, of course, in addition to the potential loss caused by under-representation of women since only 10 percent of the doctorates in science are earned by women (Harmon, 1965:53–61).

3. For economy and to avoid repetition, the abbreviation "HEP" will be used throughout to indicate "high energy physics" or "high energy physicist(s)," depending upon the context.

4. Italics added. Meier (1951:96–97), a social scientist, states: "During the early nineteenth century the stature of pure science as an intellectual discipline fully equal to fine arts and classics developed simultaneously in most European countries. In recent times, however, in all but a few of them the tradition has been overwhelmed by war and commercialism. In Britain the tradition remains strong, and strenuous attempts are still made to refine and purify the sciences—especially in the universities which are responsible for their propagation. Many scholars still intentionally select subjects which in their judgment have no foreseeable application."

5. Only four of the 203 HEP studied have no Ph.D. degree. Of these, three

have published research that probably would be acceptable as a thesis; traditionally, course work past the undergraduate degree has not been a requirement for a doctorate.

6. The most informative reviews and extensive bibliographies dealing with most of the writings about science and scientists are found in a few sources. Among the best are Barber (1956; 1959; 1962), Barber and Hirsch (1962), Kaplan (1964; 1965), Hagstrom (1965), Storer (1966), and Hirsch (1968).

While research in Britain has not focused on the same problems, some data on British scientists are found in Halsey and Trow (1967; 1971), Box and Cotgrove (1966; 1968), Ford and Box (1967a; 1967b), Hudson (1967), Hutchings (1967), and Rudd (1968).

2 The Organization of Basic Research

1. There has been some change in the organization of research away from the almost completely autonomous research councils that have allocated funds. Some of the money is to be spent by government departments on a contract basis, buying specific research from presumably the same researchers who formerly obtained funds from the research councils. As of autumn 1972, the Science Research Council is not among the affected councils.

2. Two journals, *Minerva* and *Universities Quarterly,* regularly publish articles, letters, and exchanges about the actual or threatened encroachment upon academic freedom that results because universities are funded by the central government through the UGC.

3. At present there is one important social consequence of decentralization in the United States. Militant students are able to point out contract research support from the Department of Defense as a sellout by university professors. On or around 15 January 1969, the student newspaper of Boston University devoted the front page to a list of faculty members. The caption was to the effect, "What do these professors have in common?" The story inside "exposed" their DOD contract research. Some were theoretical physicists, receiving support for research in elementary particles—a far cry from immediate DOD defense needs. There are, of course, two sides to this problem. First, many students are simply ignorant of the fact that much of the DOD-supported research has no connection at all with weapons. And yet if professors receive DOD funds, students are at least bright enough to suspect that in some cases at least a DOD-supported professor may be hesitant to criticize DOD policies (which some students oppose, believing their professors should as well).

4. In the year ending 30 June 1968 the United States spent approximately 160 million dollars on high energy physics—about four times that of Great Britain (High Energy Physics Advisory Panel, 1968). The United States, of course, has about four times the population and more than four times the Gross National Product of Great Britain. Since high energy physics is essentially economic consumption rather than investment—at the moment at least—the British support is dearer than the United States support.

5. The RHEL accelerator is called "Nimrod" (from Gen. 10: 8: "And Cush begat Nimrod: he began to be a mighty one in the earth"). The DNPL accelerator is called "Nina" (from National Institute Northern Accelerator). The SRC was formed in 1965 to assume the obligations of the original National Institute for Research in Nuclear Science.

6. For a complete enumeration by institution and amount see University Grants Committee (1968).

7. Halsey (1962:93) shows that government grants as a percentage of total university expenditures rose from 35.8 percent in 1938–39 to 72.2 percent in 1959–60 while funds from endowments fell from 15.4 percent to 2.9 percent in the same period.

3 Research in High Energy Physics

1. The information for chap. 3 was obtained primarily through discussion with British HEP before or after interviewing them for this study. Before going to Britain, there were discussions with American and British scientists at Brookhaven National Laboratory, the joint Harvard-Massachusetts Institute of Technology laboratory (Cambridge Electronic Accelerator), Columbia University, Yale University, Brown University, and the University of Pennsylvania (whose scientists not only use Brookhaven National Laboratory but who used to have their own Princeton-Pennsylvania accelerator). I am grateful to a scientist in the study, who prefers to remain anonymous, who read this chapter and offered suggestions.

2. Quoted by Swatez (1966:33). Swatez's report involving research at the Lawrence Radiation Laboratory at Berkeley contains interesting historical material regarding the development of high energy physics; for greater detail of some of the points mentioned in this chapter the interested reader is referred to that source. A chronological history of nuclear physics may be found in a mimeographed publication compiled by the Center for History and Philosophy of Physics (1968).

3. Greenberg's (1967: 97–98) comments on the effect of the war on American physics are of interest: "World War II shattered the long-dominant European scientific community. . . . But the historically penurious American scientific community, though almost wholly diverted to war work, was invigorated and enriched by the war. Europe lost Fermi and Einstein; the United States gained them. Many of the workshops of European science were weakened or destroyed by war, but the workshops of American science were strengthened by the forced-draft application of science and technology to war. After the liberation, while European science faced the task of reconstruction, the American scientific community was going through what was probably the first of the postwar revolutions of rising expectations. The war demonstrated not only that big organization, big equipment, and generous funding were compatible with the creative process, but that with war-born instruments and technology ready to be applied to basic research, bigness had become indispensable in many fields of research, especially physics. In the prevailing understanding, technology was the offspring of fundamental knowledge—knowledge led to devices and gadgets; but, in fact, the instruments of war that were developed through fundamental knowledge could just as well serve to produce new fundamental knowledge. . . . Now the vacuum and electronics technology that had been developed for radar opened possibilities for a new generation of particle accelerators to explore deeper into the atom."

4. For a readable history of accelerators as machines and tools for nuclear physics see Wilson and Littauer (1960).

5. For example, the Nimrod accelerator at the Rutherford High Energy Laboratory is designed to produce one trillion protons per pulse of the machine, with twenty-eight pulses a minute (Science Research Council, n.d.-a).

6. These data exemplify two things, the interchangeability of scientists and, to a lesser extent, the "brain drain."

7. See Libbey and Zaltman (1967) for some problems faced by theorists in such desolate environments.

8. For nontechnical literature, the interested reader is referred to Frisch and Thorndike (1964), Livingston (1963), Yang (1962), and Yuan (1965).

9. See Shutt (1967) for papers by various experts on the development and types of bubble chambers.

10. In 1959, design work began at Brookhaven National Laboratory in New York on an 80-inch bubble chamber. The whole project to completion and including building facilities required 250 man-years and cost almost $6 million. Between June 1963 and July 1966 3.3 million pictures were taken for twenty-nine experiments of scientists at seven universities and laboratories. It takes fifty people to operate and maintain the chamber continuously. See Shutt (1966) for an interesting account of this chamber and for citations to materials written for lay understanding.

11. With very few exceptions, every quotation taken from interview data is a direct quotation because most interviews were tape-recorded (these are *verbatim* transcriptions). I corrected grammatical errors and deleted proper names and places that might reveal individual identities or reflect upon persons and institutions.

4 The Reward System

1. Scientific research involves activity, of course, which is an important part of life—people enjoy doing something. Out of the 203 HEP, 99 said there was very little or no pressure on them to do research. Realizing that the answer to the question of "Why do you do research?" can only be taken as data on "response to the question" rather than why they really do research, the responses were: personally enjoyable (47%), career reasons such as promotions and conditions of appointment (28%), and both (25%).

2. See Cartter (1966) for citations and discussions of studies of American university prestige. A strong correlation exists between several objective and subjective measures of university prestige. For example, faculty salaries, size of library, quality of faculty, and student abilities are all interrelated and form the bases for opinions about prestige. The wealth at the disposal of the university is yet another important element related to all these factors. Even with equivalent funds, however, no one seriously suggests that all differences between universities would be overcome. Because the "university system" in the United States has private and public constituents and is so diverse, it may not be valid to call it a system. (Parsons and Platt [1969] argue that it is in fact a system.)

3. Another factor may be recruiting practices. Crane (1969) has shown, with Berelson's (1960) data on American academics, that social class is positively related to the prestige of academics' affiliations even when controlling for the prestige of the university granting the doctorate.

4. Twelve HEP were classified on the basis of the mothers' occupations because of the fathers' absence through death or divorce. If there was a stepfather in the home when the scientist was twelve, his occupation was used. Occupations were classified according to the five-group classification of the Registrar General (General Registrer Office, 1966). Class I, the top category, is comprised of professional occupations such as university teachers and judges.

Class II represents intermediate occupations such as secondary-school teachers and nurses. Class III is comprised of skilled occupations—office machine and telegraph operators, etc. Class IV, semiskilled workers, includes restaurant waiters and deliverymen. Class V is comprised of manual workers such as porters and office cleaners.

5. Berelson (1960:178) queried faculties in American graduate schools and reported that "According to the faculty, the student selects his own topic in this many cases: Physical sciences 2% . . ." The accurate interpretation of these figures differs from Berelson's report. What he actually meant is that 2 percent of the faculty in the physical sciences report that students independently select their own topics—a very different statement from faculties saying that 2 percent of all students select their own topics. In the correct interpretation of these data there is still sufficient evidence that science students seldom choose their own topics. The faculties' statements about these matters may be biased sources of information because Berelson also surveyed recent doctorates who indicated themselves more involved in the decision than the faculty reported them as being.

6. I am grateful to the Science Research Council for privately furnishing these data on funding.

7. For the HEP, publication of books is highly correlated to productivity; 72 percent (out of 18) of those who published books were high in productivity whereas only 30 percent who did not publish books were high (chi-square=11.45; 1 df, prob. <.001; gamma=.72).

8. In sociology, for example, the "visiting sociologist" program sponsored by the American Sociological Association and funded by a national agency was to provide students at institutions having relatively few well-known sociologists an opportunity to see what a "real" sociologist is like. The visiting sociologist enjoys a certain status for being a messenger to these backwaters.

9. To test this hypothesis, one could look at the relative frequency persons chosen for such tasks send alternates (if possible) as compared to the frequency of sending alternates (when possible) to sit on other types of committees.

10. See Crane (1964) for questions dealing with the influence of the adviser. The prestige of the thesis adviser involved categorization based on having "high" or "low" recognition in the present study (if he was in the present study), or meeting one of several criteria such as being a Fellow of the Royal Society or being listed in *Who's Who in Science in Europe* (1967). The criteria of inclusion are not explicit, but a large proportion of the "high" recognition scientists in the present study were listed in the book.

11. Interestingly enough, the Coles (1968:400) found a correlation of .24 between rank of department and quantity of publications for their sample of one hundred and twenty American physicists. Eminent scientists are overrepresented in their study; therefore, high-prestige departments may be assumed to be overrepresented.

12. The responses of British scientists in America may have resulted from guilt feelings, and therefore are rationalizations, because of the considerable publicity in various media to the so-called brain drain. Of course, in their own careers it could be true since their experiences before coming to the United States were limited because of the short time they had been scientists. One experimentalist was asked about younger men participating in important matters, and he replied: "Higher level committees in England are the old men—the professors—they make the real decisions. It's one of the unfortunate features. It's rather better here. You find that the people who make the decisions are

usually younger than in England. I think this is an advantage in this country. In England, it seems to me there's a tendency that if you're going to start off a new project, a rather senior person would have to take charge. In England if there was some kind of work, it would be inconceivable that I should take charge of it. Over here I was able to do it, and it turned out successfully. I could have done the same in England, but the attitudes are such that you aren't able to."

13. Quality of research has not been measured directly in this study because quality is highly correlated with quantity (see Cole and Cole, 1967); and without expending a large amount of research time or money, one may assume that, on the average, high producers produce high quality work.

5 Competition in Science

1. For a discussion of collaboration see Hagstrom (1965: chap. 3; 1967).
2. I am grateful to Professor Edward Shils for bringing this novel to my attention.
3. See Merton (1957:537–49) and Barber (1962) for citations to relevant publications and for a discussion of this aspect of discrimination.
4. The same explanation could be used to account for the different achievements between nations in elementary particle physics research. There would be one problem in using Ben-David's method. He did not standardize his main indices (number of medical discoveries and the number of discoverers) for population differences, and he argued convincingly that it was not necessary. Modern scientific research, and especially experimental elementary particle physics research, requires an enormous amount of research funds, so an important factor in comparing nations would be their relative wealth.
5. That is, pure or basic scientists cannot do this. Scientific inventors may work in a system where "invention is the mother of necessity." Scientists who develop weapon systems, etc., may also—at least so they are commonly accused—produce knowledge that then apparently must be translated into actual "hardware."
6. Using 264 instances of multiple discoveries, Merton (1965:125) reports trends between early and recent periods. Before 1700, 92 percent (out of 36 cases) of the multiple discoveries resulted in strenuous conflicts over priority; through the 1700s and up through the middle of the 1800s, the percentage involving priority conflicts was about 72. The second half of the nineteenth century shows a drop to 59 percent, and for the first half of the twentieth century the percentage drops to 33 percent. "It may be that scientists are becoming more fully aware that with vastly enlarged numbers at work in each field of science, a discovery is apt to be made by others as well as by themselves." This suggests that multiple discoveries may be made by more than two or more people, and carrying out a priority dispute with several people could be difficult. Since journals record the time and day papers arrive (see Hagstrom, 1967) there may now be a more institutionalized method of deciding who is in fact first.
7. To be completely accurate, Hagstrom suggested that chemistry would also fall between physics and the formal sciences. He did not present chemistry or physics separately in the tables showing the results, so it can only be assumed that he correctly included both of them in the category "physical sciences."
8. Hagstrom (1967:75) suggested that concern is a function both of preva-

lence and severity of competition, but that he would assume that severity is the more important component.

9. Using concern about anticipation as a measure of severity, as in the early, small sample, the distribution for severity showed the physical sciences most severe, the formal sciences in the middle, and the biological sciences least severe.

Concern may not be an appropriate measure of severity since it calls for subjective appraisal, and Hagstrom (1967:7) admits this. Concern brings into play the degree to which a person *cares* about being anticipated. Hagstrom (1967:7) suggests that young scientists may fear anticipation because of their lower status and the fact that the experience of anticipation might be more damaging and because they do not know what might result from anticipation.

10. An attempt was made to get the opinions of HEP on the question of the status of theory. They were asked: "Is there a general body of theoretical principles about which most scientists working in your area agree?" Of the 161 scientists asked, 39 percent said "yes" and that there had *not* been a period in the recent past when this agreement did not exist; 27 percent said "yes" but that there *had been* a period in the recent past where the agreement was not present; 28 percent said there was no agreement on theoretical principles either now or in the past; and 6 percent said there was no present agreement, but there had been agreement in the recent past. The results are extremely tentative. The concept of theory or "principles" is not specific, even to physicists, and it was difficult to explain exactly what was being asked. The question seemed to call for answers about their specific problem area rather than the whole field of elementary particles.

11. Some figures will illustrate the size of the research tools used in HEP, with even larger machines now being used but not involved at the time the interviews were made. A comparison between the Rutherford High Energy Laboratory accelerator (RHEL) in Britain and the Brookhaven National Laboratory accelerator (BNL) in the United States (with the BNL figures in parentheses following the RHEL) shows that for RHEL, the operating energy is 7 BeV (30), the diameter of the accelerator is 155 feet (843), the number of magnets is 336 (240), the total weight of the magnets is 7,000 tons (4,400), the number of pulses per minute is 28 (25), the number of protons per pulse is 2×10^{12} (2×10^{11}), and the accelerated particles travel 100,000 miles (180,000) before colliding with a target.

Operating at 10.5 BeV, the BNL accelerator makes 75 pulses each minute, thereby producing many more protons per minute than the RHEL accelerator, and this increases many times the probability of producing an interesting interaction and provides advantages to scientists doing research at lower energies on more powerful accelerators.

12. Such opinions are probably based on out-of-date facts. One very eminent British scientist, possibly more informed than other scientists, said (in September 1967) in answer to the question of British financial research support: "Well, nowadays there's really nothing they couldn't do." At about this same time (1968), a panel on HEP was preparing a report showing that the United States was falling behind western Europe in financial support for high energy physics.

13. The political reason for locating the Nina accelerator at Daresbury Nuclear Physics Laboratory in the north of England to serve the northern and Scottish universities is denied by no one. An eminent immigrant to the United States said: "I think I'm partly responsible for the [northern machine]. I left for I was getting no response or help and sort of pushed things along by leaving.

They are all scared about the 'brain drain.' About a ——— [time period] after I left, they got the machinery."

One comment about this informant is in order. When the tape recorder was turned off and the interviewer started to leave, he said: "I'm glad you didn't ask me any questions about why I left England. I'm always getting questionnaires asking me to check off my reasons for leaving England; more money, etc. I don't really know, and if I did it's none of their business. I just throw them away."

This response is perhaps explained by his apparent guilt about immigration because he said that when he left England as a "brain drain loss" people were individually blamed whereas now it's simply part of a phenomenon.

14. The concept "beam" refers to the type of particles that is caused to hit whatever target is being used. Protons were the initial particle, but interactions with other particles were caused to produce a secondary "beam" of K^- particles that was directed into the hydrogen target, being the liquid hydrogen contained in the bubble chamber.

15. A bubble chamber group of significant strength may do several "experiments" or film analyses simultaneously. For any one experiment, three or four scientists may work together, each of whom might be involved as one of three or four scientists working on a different experiment. Many permutations are therefore possible.

16. The letters, of course, were published in sequence, with the American's first, not "side by side." It is significant that the loser stresses "side by side" while the winner stresses "ours first."

6 Competition in High Energy Physics

1. Having been anticipated does not affect one's recognition (table not shown, chi-square significant at .50, gamma=.20). Scientists with high recognition are more likely to be anticipated, but the differences between high and low recognition are not great and are almost exactly the same distribution as that for high and low productivity.

2. This raises an interesting question of what constitutes anticipation. If work is exactly the same and published at the same time, it is multiple discovery; there is clear anticipation only if one is first. If work overlaps, it probably signifies that one scientist has published more detailed work rather than that he has hit upon an entirely new discovery. In fact, controlling for field of research, experimental physics and the biological sciences were the most likely to have published their research after being anticipated (Hagstrom, 1967:8). The biological sciences are certainly more experimental than theoretical.

3. See Merton (1968) for a discussion of the biblical statement in Matt. 25:29: "For unto every one that hath shall be given, and he shall have abundance: but from him that hath not shall be taken away even that which he hath."

4. The use of the concept "formal publication" implies that other forms of publication are possible. For evidence that preprints as a type of publication cause some problem among scientists see Libbey and Zaltman (1967:B1-B3).

5. Hagstrom (1967:127) reports that 54 percent of both theoretical and experimental physicists in the United States did *not* know of an instance when they should have been cited but were not.

6. Fear of the theft of ideas may be more prevalent in the United States.

Orlans (1962:193–94) states: "Some successful as well as unsuccessful applicants [to federal agencies for research funds] expressed apprehension that their ideas might be exploited by agency staff, members of reviewing panels, or authorities to whom a proposal is sent for evaluation. Occasional incidents have lent some substance to this concern, although most agency staff and panel members undoubtedly behave with discretion. The fear of being beaten to a good experiment or of not receiving credit for a good idea is more pronounced in America than in European countries where, since every scientist in a particular field knows every other scientist and what he is doing, there is less pressure to publish and less opportunity to capitalize on someone else's ability. Some experienced American scientists take such steps as they feel necessary to protect their priority, withholding from proposals and early publications critical information that would permit too early replication of their work."

7. The possibility of this event involving a Japanese journal is indirectly upheld by Zaltman (1968:37) who found the Japanese theoretical high energy physics community to be relatively isolated vis-à-vis other subgroups of the international community of theoretical high energy physicists. On a measure involving nominations of informal information contacts, the Japanese (one of the five largest groups of theoretical high energy physicists) were the least likely (6 percent) to name theorists whose country of employment and nationality were not Japanese. This contrasts with West Germany's 46 percent, which was the highest percentage.

8. Hagstrom (1967:10) found that the percentage differences for concern about anticipation between scientists who had never been anticipated and those anticipated three times or more ranged from 22 to 44 percent, depending on the year of the scientists' doctorates. There is a complex interaction effect between experience of anticipation and professional age so that the magnitude of the percentage differences is not related linearly to year of doctorate.

Among the HEP, those with no anticipation, one anticipation, and two (plus) anticipations were, progressively, more likely to report the probability was high that their present research will be anticipated (chi-square=8.25, 4 df., prob. <.10; gamma=.23).

9. Competition has been measured using the percentage of HEP and physicists (in Hagstrom's sample) who have experienced anticipation, but anticipation is a zero sum phenomenon: for every anticipation, someone has been *anticipated* and someone has been the *anticipator.* To measure competition, then, a researcher should know both who has been anticipated and the number of times each scientist has anticipated others. The addition of these two proportions, for example, would give more clearly the extent of competition in a field. It is not always possible for scientists to know whom they have anticipated—especially if the anticipated scientists did not publish as a result.

In addition to considering anticipators in a measure of competition, consideration should be given to the reasons why scientists have not been anticipated. If being anticipated is an index of competition, the conditions that kept scientists from being anticipated might also indicate competition. One scientist volunteered that he had never been anticipated because of his group's competence. Another said he had not been anticipated because of hard work. "I think it is a case of working eighteen hours a day." The 66 HEP not anticipated gave these reasons: luck (8 percent); not in a topical area (44 percent); never been in direct competition such as a race (35 percent); and were anticipators themselves because they won a race (14 percent). Competition, therefore, may be present when there is no indication of it through self-reported anticipations.

When competition becomes too stiff in some research areas, scientists switch to others (Hagstrom 1967:14) where they are able to make discoveries and be rewarded. Among Hagstrom's sample, physicists (8–9 percent) were the most likely to have already changed specialties because of "too many scientists conducting research" in their area. More than twice that proportion had considered doing so but not yet changed. An improved competition indicator might include this "change" factor.

Among the HEP, 46 percent (of 196) had thought about leaving elementary particle research because of frustration and doubts about the future (41 percent) and for other reasons (5 percent) such as returning to a former research specialty. That percentage is much greater than the percentages reported by Hagstrom. Most of the interviews were conducted after the British government had devalued the pound in November 1967. Economies in all government spending were one of the main topics of conversation at that time. High energy physics research is very expensive, and there are constant allegations by public speakers that not enough scientists (in all disciplines) are interested in the kinds of research that would benefit the country's economy. HEP especially are not contributing to the economy but are a big drain on the economy (relative to other sciences). Some of the HEP seemed rather conscience-stricken. Also, doubts about the future came from the well-known fact that the next generation of particle accelerators must be an international endeavor, in contrast to the present situation where there are both national machines and the larger international machine at CERN. It is understandable that HEP might be concerned about the future of research in their specialty if they are forced to do almost all of their research at a foreign laboratory and at the same time are required to teach undergraduates at their home university.

7 Communication in the High Energy Physics Community

1. One original objective in this research was to use Price's suggestion (1963:84) that it might be possible to trace the networks of scientists through their personal preprint mailing lists by taking each individual's list and establishing a sender-receiver matrix. Unfortunately, most HEP use laboratory services where a master list is automatically used to distribute preprints. (Some HEP had lists, but most did not.) Names are added to the master list when the receiver requests his name be added or when individual staff members suggest the names of potential receivers. Thus mechanical distribution represents the wishes of no one person. It is more efficient to send preprints regularly to names on a master list rather than single out individual recipients. Some preprint distributions, therefore, are rather like journal distribution.

Even staff secretaries realize the function of preprints. One librarian responsible for sending out preprints for a university department was very cooperative. She allowed me to see her list and told me how many additions she had made since taking over the responsibility. She said, "It's very important to get their preprints out very quickly, to help their reputations, you know."

2. A third function of informal communication is simply to keep up with the interesting gossip that is prevalent in any human group; who has made off with whose wife; who is getting sacked; and who applied for but failed to get a professorship, etc.

3. This is a type of verbal communication, but it is passive rather than active. It is similar to listening to someone read aloud a preprint that will soon be published.

4. Controlling for professional age in considering the relationship between most important communication channel and rank and controlling for rank in considering the relationship between professional age and most important communication are not possible because the controls result in very small numbers in most cells.

5. The mean rank for *Physical Review Letters* was 1.51; for *Physical Review,* 2.44; for *Physics Letters,* 2.65; for *Nuovo Cimento,* 3.40. British journals for physics research include *Nature,* a weekly letter journal for all fields of science (at the time of the interviews), *Proceedings of the Physical Society,* published by the Institute of Physics and the Physical Society, *Proceedings of the Royal Society,* and *Proceedings of the Cambridge Philosophical Society.*

The ranking of journals by British HEP does not represent their journals of publication. The actual publications of the 88 HEP who had lists of their research showed 997 titles, published as follows: *Physical Review Letters,* 5 percent; *Physical Review,* 12 percent; *Physics Letters,* 8 percent; *Nuovo Cimento,* 16 percent; *Proceedings of the Physical Society,* 10 percent; *Nuclear Physics,* 9 percent; and, twenty other journals, 40 percent. These data are tentative because the list is not a systematic sample, and the titles may be listed in the bibliographies of several scientists because of joint publications. Whether that overestimates publications in core journals is a guess. Perhaps HEP who publish in core journals are more likely to maintain their bibliographies, but no data are available to substantiate that explanation. Not all titles are research results on elementary particles since some HEP were nuclear structure physicists before there was high energy, and still fewer come from other research specialties. The appropriate method of obtaining exact journal distributions is to obtain data directly from the first page of every published paper, a task that would require several man-months, and the data obtained probably would not differ greatly from these distributions.

6. Libbey and Zaltman's sample (1967:41) of 977 theoretical high energy physicists found that 94 percent regularly scanned both *Physical Review* and *Physical Review Letters,* showing the world-wide influence of American theoretical physics.

7. Six out of the sixteen HEP in the invisible college made reciprocal choices. Other choices of the invisible college were eight scientists who made sixteen choices themselves, ten of which were in the invisible college.

8. If all types of connections including collaboration (either past or present) rather than only communication ties are considered, the connectivity scores would be larger. Since many collaborations are not voluntary but forced because of administrative directives from laboratories, they form a different kind of tie. Not to include collaborative ties overlooks both the voluntary collaborations that do exist and the communication that might result were the individuals not involved in a group collaboration. The purpose here is to discuss purely voluntary communication that, of course, will be conservative because of these exclusions.

8 Summary and Conclusions

1. One advantage of Oxbridge has been, of course, its social life, perhaps as important as its intellectual superiority among British universities. Some of the advantages that are usually pointed out about Oxbridge—entrance into the civil service, for example—are said to result mainly from the influential people

one meets. Examinations of a general nature are much easier for students who have been associated in college residences where they are exposed to students reading for degrees in every subject. This contrasts to a university where only a small percentage of the students live in university housing and contact with other students is predominantly in one's own department.

2. This discussion is obviously indebted to Professor Joseph Ben-David who has been concerned with one or another form of competition in the several papers listed in the bibliography.

3. The HEP community is rather large compared to some groups. See Watson (1969) for an example of a small group of researchers involved in the discovery of the structure of DNA and the consequent distribution of rewards.

4. Of the pure sciences in the United States, high energy physics is the most centralized and does have the most explicit science policy since the major funding of accelerators is from the Atomic Energy Commission. High energy physics is by far the most expensive. In Britain, in contrast, most disciplines have a central policy that derives from the same committees and departments as the policy for the more expensive high energy physics.

Bibliography

Adams, J. B.
1965 "CERN: The European Organization for Nuclear Research." Pp. 236–61 in Sir John Cockroft (ed.), *The Organization of Research Establishments.* Cambridge: Cambridge University Press.

American Men of Science
1965–67 *A Biographical Directory.* New York: Science Press (11th ed).

Associated Universities Incorporated
n.d. "The Brookhaven Alternating Gradient Synchrotron" (brochure).

Association of Commonwealth Universities
1967 *Commonwealth Universities Yearbook.* London: Association of Commonwealth Universities.

Association of University Teachers
1965 *The Remuneration of University Teachers, 1964–1965.* London: Association of University Teachers.

Barber, Bernard
1956 "Sociology of Science: A Trend Report and Bibliography." *Current Sociology* 5:91–153.
1959 "The Sociology of Science." Pp. 215–28 in Robert Merton et al. (eds.), *Sociology Today.* New York: Basic Books.
1962 *Science and the Social Order.* New York: Collier Books (rev. ed).

Barber, Bernard, and Walter Hirsch (eds.)
1962 *The Sociology of Science.* New York: Free Press.

Barzun, Jacques
1964 *Science: The Glorious Entertainment.* New York: Harper and Row.

Ben-David, Joseph
1965a "Scientific Productivity and Academic Organization in Nineteenth Century Medicine." Pp. 39–61 in Norman Kaplan (ed.), *Science and Society.* Chicago: Rand McNally.
1965b "The Scientific Role: The Conditions of its Establishment in Europe." *Minerva* 4 (Autumn):15–54.
1968 *Fundamental Research and the Universities.* Paris: Organization for Economic Co-operation and Development.
1969 "National and International Scientific Communities." Paper prepared for IUHPS/UNESCO Study of National Scientific Communities Conference. Nairobi, Kenya (20–26 January).

Ben-David, Joseph, and Randall Collins
1966 "Social Factors in the Origins of a New Science: The Case of Psychology." *American Sociological Review* 31 (August):451–65.

Ben-David, Joseph, and Awraham Zloczower
1962 "Universities and Academic Systems in Modern Societies." *European Journal of Sociology* 3:45–84.

Berelson, Bernard
1960 *Graduate Education in the United States.* New York: McGraw-Hill.

Blalock, Hubert M.
1960 *Social Statistics.* New York: McGraw-Hill.

Box, Steven, and Stephen Cotgrove
1966 "Scientific Identity, Occupational Selection, and Role Strain." *British Journal of Sociology* 17 (March):20–28.
1968 "The Productivity of Scientists in Industrial Research Laboratories." *Sociology* 2 (May):163–72.

Brown, David
1967 *The Mobile Professors.* Washington, D.C.: American Council on Education.

Caplow, Theodore, and Reece J. McGee
1965 *The Academic Marketplace.* New York: Doubleday (first published 1958).

Cartter, Allan M.
1966 *An Assessment of Quality in Graduate Education.* Washington, D.C.: American Council on Education.

Center for History and Philosophy of Physics
1968 "Nuclear Physics History Chronology: Preliminary Version." New York: American Institute of Physics (mimeographed).

Cockroft, Sir John (ed.)
1965 *The Organization of Research Establishments.* Cambridge: Cambridge University Press.

Cole, G. D. H.
1955 *Studies in Class Structure.* London: Routledge and Kegan Paul.

Cole, Stephen, and Jonathan R. Cole
1967 "Scientific Output and Recognition: A Study in the Operation of the Reward System in Science." *American Sociological Review* 32 (June):377–90.
1968 "Visibility and the Structural Bases of Awareness of Scientific Research." *American Sociological Review* 33 (June):397–413.

Coleman, J. S.
1964 *Introduction to Mathematical Sociology.* New York: Free Press.

Collins, Randall
1967 "Competition and Social Control in Science." Paper presented at the American Sociological Association, San Francisco (August) (mimeographed).
1968 "Competition and Social Control in Science: An Essay in Theory Construction." *Sociology of Education* 41 (Spring):123–40.

Committee on Higher Education
1963 *Higher Education.* London: H.M.S.O.

Consolazio, William V.
1967 *The Dynamics of Academic Science.* Washington, D.C.: National Science Foundation.

Council for Scientific Policy
1966 *Report on Science Policy.* London: H.M.S.O.
1967 *Second Report on Science Policy.* London: H.M.S.O.
1968 *Enquiry into the Flow of Candidates in Science and Technology into Higher Education.* London: H.M.S.O.

Crane, Diana
1964 "The Environment of Discovery: A Study of Academic Research Interests and Their Setting" (unpublished Ph.D. dissertation, Columbia University).
1965 "Scientists at Major and Minor Universities: A Study

of Productivity and Recognition." *American Sociological Review* 30 (October):699–714.

1967 "The Gatekeepers of Science: Some Factors Affecting the Selection of Articles for Scientific Journals." *American Sociologist* 2 (November):195–201.

1969a "Fashion in Science: Does It Exist?" *Social Problems* 16 (Spring):433–41.

1969b "Social Class Origin and Academic Success: The Influence of Two Stratification Systems on Academic Careers." *Sociology of Education* 42 (Winter):1–17.

1969c "Social Structure in a Group of Scientists: A Test of the 'Invisible College' Hypothesis." *American Sociological Review* 34 (June):335–52.

1972 *Invisible Colleges.* Chicago: The University of Chicago Press.

Crowther, J. G.

1952 *British Scientists of the Twentieth Century.* London: Routledge and Kegan Paul.

Demerath, Nicholas, Richard W. Stephen, and R. Robb Taylor

1967 *Power, Presidents, and Professors.* New York: Basic Books.

Department of Education and Science

1968 *The Proposed 300 GeV Accelerator.* London: H.M.S.O.

Department of Education and Science and the British Council

1967 *Scientific Research in British Colleges and Universities.* London: H.M.S.O.

Directory of British Scientists

1966 London: E. Benn (2 vols.)

Edson, Lee

1967 "Two Men in Search of the Quark." *New York Times Magazine* (8 October):60+.

Fletcher, F. T. H.

1961 "United Kingdom." Pp. 165–78 in Richard Shryock (ed.), *The Status of University Teachers.* Paris: UNESCO.

Floud, Jean, and A. H. Halsey

1961 "English Secondary Schools and the Supply of Labor." Pp. 80–92 in A. H. Halsey, Jean Floud, and C. Arnold Anderson (eds.), *Education, Economy, and Society.* New York: Free Press.

Ford, Julienne, and Steven Box

1967a "Commitment to Science: A Solution to Student Marginality?" *Sociology* 1 (September):225–38.

1967b "Sociological Theory and Occupational Choice." *Sociological Review* 15 (November):287–99.

Ford, K. W.
1963 *The World of Elementary Particles.* New York: Blaisdell.

Frisch, D. H., and A. M. Thorndike
1964 *Elementary Particles.* Princeton: Van Nostrand.

Furneaux, W. D.
1961 *The Chosen Few.* London: Oxford University Press.

Garvey, William D., and Belver C. Griffith
1966 "Studies of Social Innovations in Scientific Communication in Psychology." *American Psychologist* 21 (November):1019–36.

Gaston, Jerry
1969 "Big Science in Britain: A Sociological Study of the High Energy Physics Community" (unpublished Ph.D. dissertation, Yale University).

General Register Office
1966 *Classification of Occupations.* London: H.M.S.O.

Glass, Bentley, and Sharon H. Norwood
1959 "How Scientists Actually Learn of Work Important to Them." Pp. 195–97 in *Proceedings of the International Conference on Scientific Information* 1. Washington, D.C.: National Academy of Sciences, National Research Council (2 vols).

Gold, David
1969 "Statistical Tests and Substantive Significance." *American Sociologist* 4 (February): 42–46.

Goldberger, M. L.
1968 "Tribute to Bethe." Review of R. E. Marshak's *Perspectives in Modern Physics. Science* 160 (May):666–67.

Greenberg, Daniel S.
1967 *The Politics of Pure Science.* New York: New American Library.

Gross, Edward
1968 "Universities as Organizations: A Research Approach." *American Sociological Review* 33 (August): 518–44.

Gruber, Howard E., Glenn Terrell, and Michael Wertheimer
1963 *Contemporary Approaches to Creative Thinking.* New York: Atherton Press.

Hagstrom, Warren O.
1965 *The Scientific Community*. New York: Basic Books.
1966 "Competition in Science." Paper prepared for the American Sociological Association, Miami Beach (August) (mimeographed).
1967 "Competition and Teamwork in Science." Final Report to the National Science Foundation (mimeographed).
1968 "Departmental Prestige and Scientific Productivity." Paper prepared for the American Sociological Association, Boston (August) (mimeographed).

Halsey, A. H.
1961 (Ed.) *Ability and Educational Opportunity*. Paris: O.E.C.D.
1962 "British Universities." *European Journal of Sociology* 3:85–101.

Halsey, A. H., and Martin Trow
1967 *A Study of the British University Teachers*. Final Report to the U.S. Office of Education (mimeographed).
1971 *The British Academics*. Cambridge, Mass.: Harvard University Press.

Hargens, Lowell L., and Warren O. Hagstrom
1967 "Sponsored and Contest Mobility of American Academic Scientists." *Sociology of Education* 40 (Winter):24–38.

Harmon, Lindsey R.
1965 *Profiles of Ph.D.'s in the Sciences*. Washington D.C.: National Academy of Sciences, National Research Council.

High Energy Physics Advisory Panel
1968 "The Status and Problems of High-Energy Physics Today." *Science* 161 (July):11–19.

Hirsch, Walter
1968 *Scientists in American Society*. New York: Random House.

Hiscocks, E. S.
1959 "Organization of Science in the United Kingdom." *Science* 129 (March):689–93.

Hornig, Donald F.
1969 "United States Science Policy: Its Health and Future Direction." *Science* 163 (February):523–28.

Hudson, Liam
1963 "The Relation of Psychological Test Scores to

Academic Bias." *British Journal of Educational Psychology* 23 (June):120–31.

1967 "The Stereotypical Scientist." *Nature* 213 (January 21):228–29.

Hutchings, Donald

1967 *The Science Undergraduate.* Oxford: Oxford University Department of Education.

Jungk, Robert

1968 *The Big Machine.* New York: Scribner.

Kaplan, Norman

1964 "The Sociology of Science." Pp. 852–81 in Robert E. Faris (ed.), *Handbook of Modern Sociology.* Chicago: Rand McNally.

1965 (Ed.) *Science and Society.* Chicago: Rand McNally.

Klaw, Spencer

1968 *The New Brahmins.* New York: Morrow.

Kornhauser, William

1962 *Scientists in Industry.* Berkeley: University of California Press.

Kuhn, Thomas S.

1962 *The Structure of Scientific Revolutions.* Chicago: University of Chicago Press.

Larsen, Egon

1962 *The Cavendish Laboratory.* London: Edmund Ward.

Lazarsfeld, Paul F., and Wagner Thielans, Jr.

1958 *The Academic Mind.* Glencoe: Free Press.

Lederman, L. M.

1969 "A Great Collaboration." Review of Robert Jungk's *The Big Machine. Science* 164 (April):169.

Libbey, Miles A., and Gerald Zaltman

1967 *The Role and Distribution of Written Informal Communication in Theoretical High Energy Physics.* New York: American Institute of Physics.

Livingston, M. S.

1963 *High Energy Accelerators.* New York: Interscience.

Marcson, Simon

1960 *The Scientist in American Industry.* Princeton: Industrial Relations Section, Princeton University.

1962 "Decision-Making in a University Physics Department." *American Behavioral Scientist* 6 (December): 37–38.

McClelland, David C.

1963 "On the Psychodynamics of Creative Physical

Scientists." Pp. 141–75 in Gruber et al., *Contemporary Approaches to Creative Thinking.* New York: Atherton Press.

Meier, R. L.
1951 "Research as a Social Process: Social Status, Specialism and Technological Advance in Great Britain." *British Journal of Sociology* 2:91–104.

Menzel, Herbert
1962 "Planned and Unplanned Scientific Communication." Pp. 417–41 in Bernard Barber and Walter Hirsch (eds.), *The Sociology of Science.* New York: Free Press.

Merton, Robert K.
1957 *Social Theory and Social Structure.* Rev. ed. New York: Free Press (first published 1949).
1961 "Singletons and Multiples in Scientific Discovery: A Chapter in the Sociology of Science." Pp. 470–86 in *Proceedings of the American Philosophical Society* 105 (October).
1962 "Priorities in Scientific Discovery: A Chapter in the Sociology of Science." Pp. 447–85 in Bernard Barber and Walter Hirsch (eds.), *Sociology of Science.* New York: Free Press (first published 1957).
1963 "Resistance to the Systematic Study of Multiple Discoveries in Science." *European Journal of Sociology* 4:237–82.
1965 "The Ambivalence of Scientists." Pp. 112–32 in Norman Kaplan (ed.), *Science and Society.* Chicago: Rand McNally.
1968 "The Matthew Effect in Science." *Science* 159 (January):56–63.

Mullins, Nicholas C.
1968a "The Distribution of Social and Cultural Properties In Informal Communication Networks Among Biological Scientists." *American Sociological Review* 33 (October):786–97.
1968b "Social Origins of an Invisible College: The Phage Group." Revision of a paper read at the American Sociological Association Convention, Boston (August) (mimeographed).

Orlans, Harold
1962 *The Effects of Federal Programs on Higher Education.* Washington, D.C.: Brookings Institution.

Orth, Charles D., III
1965 "The Optimum Climate for Industrial Research." Pp. 194–210 in Norman Kaplan (ed.), *Science and Society*. Chicago: Rand McNally.

Parsons, Talcott, and Gerald M. Platt
1968 "Considerations on the American Academic System." *Minerva* 6:497–523.

Pelz, Donald C., and Frank M. Andrews
1966 *Scientists in Organizations*. New York: Wiley.

Pickavance, T. G.
1965 "The Rutherford High Energy Physics Laboratory." Pp. 215–35 in Sir John Cockroft (ed.), *The Organization of Research Establishments*. Cambridge: Cambridge University Press.

Polanyi, Michael
1947 "The Foundations of Freedom in Science." Pp. 124–32 in E. P. Wigner (ed.), *Physical Science and Human Values*, Princeton: Princeton University Press.

Powell, C. F.
1967a "The Changing Nature of Science." *Proceedings of the Ceylon Association for the Advancement of Science* 2:181–92.
1967b "Matter and the Universe." *Proceedings of the Ceylon Association for the Advancement of Science* 2:165–80.

Price, Derek J. de Solla
1961 *Science Since Babylon*. New Haven: Yale University Press.
1963 *Little Science, Big Science*. New York: Columbia University Press.

Price, Derek J. de Solla, and Donald deB. Beaver
1966 "Collaboration in an Invisible College." *American Psychologist* 21 (November):1011–18.

Reif, F.
1965 "The Competitive World of the Pure Scientist." Pp. 133–45 in Norman Kaplan (ed.), *Science and Society*. Chicago: Rand McNally.

Royal Society
1963 *Emigration of Scientists from the United Kingdom*. London: Royal Society.
1968 *Postgraduate Training in the United Kingdom: Physics*. London: Royal Society.

Rudd, Ernest
1968 "The Rate of Economic Growth, Technology and the Ph.D." *Minerva* 6:366–87.

Schwartz, Melvin
1962 "The Conflict Between Productivity and Creativity in Modern Day Physics." *American Behavioral Scientist* 6 (December):35–36.

Science Research Council
1967 *Report of the Council for the Year 1966–67.* London: H.M.S.O.
1968 *Report of the Council for the Year 1967–68.* London: H.M.S.O.
n.d.a "Rutherford Laboratory" (brochure).
n.d.b "Daresbury Nuclear Physics Laboratory" (brochure).

Shutt, R. P.
1966 "High Energy Physics with the Brookhaven 80″ Hydrogen Bubble Chamber" (brochure printed by Brookhaven National Laboratory, Upton, Long Island, New York).
1967 *Bubble and Spark Chamber: Principles and Use.* New York: Academic Press.

Snow, C. P.
1965 *The Two Cultures: And a Second Look.* Cambridge: Cambridge University Press.

Stein, Morris L.
1962 "Creativity and the Scientist." Pp. 329–43 in Bernard Barber and Walter Hirsch (eds.), *The Sociology of Science.* New York: Free Press.

Storer, Norman W.
1966 *The Social System of Science.* New York: Holt, Rinehart, and Winston.
1969 "The Internationality of Science and the Nationality of Scientists: A Theoretical Perspective of National Scientific Communities." Paper prepared for IUHPS/UNESCO Study of National Scientific Communities Conference. Nairobi, Kenya (20–26 January).

Swatez, Gerald M.
1966 *Social Organization of a University Laboratory.* Berkeley: Space Sciences Laboratory.

Trow, Martin
1962 "The Democratization of Higher Education in America." *European Journal of Sociology* 3:231–62.

Turner, Ralph H.
1961 "Models of Social Ascent Through Education: Sponsored and Contest Mobility." Pp. 121–39 in A. H. Halsey, Jean Floud, and C. Arnold Anderson (eds.), *Education, Economy, and Society.* New York: Free Press.

University Grants Committee
1967 *Annual Survey Academic Year 1965–1966 and Review of University Development 1962–1963 to 1965–1966.* London: H.M.S.O.
1968 *Annual Survey Academic Year 1966–1967.* London: H.M.S.O.

University of Oxford
n.d. *Application for the Quinquennium 1967–1972.* Oxford: Oxford University Press.

Vernon, P. E.
1957 *Secondary School Selection.* London: Methuen.

Watson, James D.
1969 *The Double Helix.* New York: New American Library (first published 1968).

Weinberg, Alvin M.
1967 *Reflections on Big Science.* Cambridge: M.I.T. Press.

Who's Who in Science In Europe
1967 London: F. Hodgson (3 vols).

Wilson, Logan
1942 *The Academic Man.* New York: Oxford University Press.

Wilson, Robert R., and Raphael Littauer
1960 *Accelerators.* London: Heinemann.

Winch, Robert F., and Donald T. Campbell
1969 "Proof? No. Evidence? Yes. The Significance of Tests of Significance." *American Sociologist* 4 (May):140–43.

Yang, C. N.
1962 *Elementary Particles–A Short History of Some Discoveries in Atomic Physics.* Princeton: Princeton University Press.

Yuan, L. C. L. (ed.)
1965 *Nature of Matter: Purposes of High Energy Physics.* Brookhaven: Brookhaven National Laboratory.

Zaltman, Gerald
1968 "Professional Recognition and Communication in

Theoretical High-Energy Physics" (unpublished Ph.D. dissertation, Johns Hopkins University).

Ziman, John
1968 *Public Knowledge.* Cambridge: Cambridge University Press.

Zuckerman, Harriet
1967 "Nobel Laureates in Science: Patterns of Productivity, Collaboration, and Authorship." *American Sociological Review* 32 (June):391–403.

Index

Accelerators, 7, 17, 23, 133; characteristics of, 80–81; probability of global, 174; sample list of, 26; size of, 80; types of, 81
Age, importance of, in Great Britain, 58
Agricultural Research Council, 15
Alvarez, Louis, 79
American Physical Society, 138
Anticipation, comparison of British and American scientists, 94; concern about, 117–19, 121, 167; effect of, 98–99; experience of, 83, 95; as measure of competition, 76. *See also* Priority disputes; Multiple discoveries
Atomic bomb, 23
Atomic Energy Authority, 13
Atomic Energy Commission, 8, 16, 23
Atom smashers. *See* Accelerators

Beaver, Donald deB., 179
Becquerel, Henri, 70
Beethoven, Ludwig van, 4, 74
Ben-David, Joseph, 39, 69, 72, 165, 168
Big Science, 17, 22, 26, 61, 79
British Academy, fellows, 40
British Museum, 15
Brookhaven National Laboratory, 26, 80, 86, 87
Bubble chamber, 26, 28, 29; film, 91; funds for, 44, 47

California: University of (Berkeley), 23, 79; University of (Los Angeles), 79
Canada, 25
Cartter, Allan, 34, 41, 164
Cavendish Laboratory, 22
CERN, 17, 26, 31, 37, 80, 84, 85, 86, 87, 91, 92, 111, 113, 116, 156, 170, 174
China, 25
Cockcroft, John, 22
Cole, Jonathan, 34, 66, 67, 145
Cole, Stephen, 34, 66, 67, 145
Coleman, J. S., 146
Collaboration: in experimental research, 140; forced, 162; problems in, 109; problems of, 82–83, 86–87. *See also* Social processes
Columbia University, 71
Communication, 168; with Americans, 169–70; conferences as, 132, 138; with counterparts, 140–45; efficiency of, 130–31, 145, 154, 156–57; with experimentalists, 132; formal, 131; informal, 131, 139–45; and invisible colleges, 160; methods of, 100, 132, 133–34, 160; and professional age, 134–38; and

rank, 134, 153; and recognition, 149; role of laboratory in, 156; and scientific productivity, 134, 149; seminar as, 132; and social organization, 130; and social structure, 157; and theorists, 132; and type of scientist, 137–38; verbal contacts as, 132
Communication gap, 143
Communication networks, 145; structural and scientific bases of, 147–58
Competition, 4, 69; in bubble chamber film, 78–79; consequences for science of, 70–72; diminution of, 173–74; disciplinary differences, 75–77; between experimentalists, 159; in experimental research, 78–81; in high energy physics, 77–83; individual, 70, 164; institutional, 69–70, 164; international, 167, 174; nature of, 3–5, 74; potential, 91; prevalence of, 75–77, 94, 105; races indicating, 83–93; severity of, 75–77, 101–5; in social sciences, 174; in theoretical research, 77–78; theorists in, 159
Connectivity score, 146; and group size, 155; and type of scientist, 147, 149–50
Cooper, William, 69
Cooperation, 69. *See also* Collaboration
Cooptation, 69
Copernicus, N., 74
Cornell University, 81
Council for Scientific Policy, 12, 13, 15, 16, 17; composition of, 14
Counters, 28
Crane, Diana, 34, 49, 58, 146, 177
Crowther, J. G., 70
Cryptomnesia, 109
Curie, Marie, 70
Curie, Pierre, 70

Daresbury Nuclear Physics Laboratory, 17, 80, 82
Darwin, Charles, 70
Defense, Department of, 8, 16; minister of, 13
DESY. *See* Accelerators, types of
Discrimination, 69

Education, organization of: in Great Britain, 7–8, 159; in the United States, 7, 159
Education and Science, Department of, 13, 15
Educational background: effect on recognition, 6; of scientists, 35–36
Einstein, Albert, 4
Elementary particle physics. *See* Physics
Encounter, 39
European Organization for Nuclear Research. *See* CERN

Fermi, Enrico, 74
France, 69
Fraud, 72, 105
Freud, Sigmund, 74

Garvey, William D., 134
Gatekeepers, 103
Gatekeeping, 49
Germany, 69, 70, 81
Glass, Bentley, 131
Griffith, Belver C., 134
Groves, General Leslie, 23

Hagstrom, Warren, 30, 34, 47, 59, 72, 74, 75, 76, 77, 79, 80, 94, 95, 96, 97, 101, 103, 104, 105, 112, 116, 117, 118, 121, 167, 177
Halsey, A. H., 35, 39, 40, 161
Hargens, Lowell, 34, 177
Henry, Joseph, 73
Hiscocks, E. S., 12
Hornig, Donald F., 71
Hutchings, Donald, 36

India, 25
Interviews, 10
Invisible college, 151, 168; criteria for, 151–52; evidence of, 153–54
Italian Physical Society, 139

Japan, 25
Johns Hopkins University, The, 79
Jones, Ernest, 74

Kuhn, Thomas, 77

Labor, division of, 59, 67–68, 160, 168, 170–72; effect of, 66–68, 171; implications of, 172
Laboratories: in study, 10

Lawrence, E. O., 23
Lawrence Radiation Laboratory, 23, 78
Lederman, L. M., 71
Libbey, Miles A., 25, 111, 132
Livingston, M. Stanley, 22
London Business School, 18

Manchester Business School, 18
Manchester Institute of Science and Technology, 18
Manhattan District, 23
Matthew effect, 103
Medical Research Council, 15
Menzel, Herbert, 131
Merton, Robert K., 5, 32, 70, 72, 73, 74, 109
Michelangelo, 74
Mobility: geographical, 7–8, 166; institutional, 7–8, 37, 165–66
Multiple discoveries, 73, 88–93

National Accelerator Laboratory, 26, 80
National Environment Research Council, 15, 16
National Institutes of Health, 8, 16
National Science Foundation, 8, 13, 16, 37
Netherlands, The, 139
New Scientist, 85
New York Times, 85, 86
Norwood, Sharon H., 131
Nuovo Cimento, 104, 110, 116, 138, 139

Omega-minus particle, discovery of, 83–88
Originality, 3–5, 7, 10, 68, 73, 74, 157, 166; problem of, 174
Overproductivity, 48, 62

Parliament, 13
Physical Review, 29, 104, 138, 139
Physical Review Letters, 85, 86, 89, 90, 106, 110, 113, 138, 139
Physicists: types of high energy experimental, 26–29; types of high energy theorists, 29–30; importance of types, 171
Physics: center of research for, 168; center of high energy research, 100; components of high energy community, 19; cosmic ray, 175; distribution of high energy research, 24–25; high energy experimental, 26–29, 170–71; high energy theoretical, 29–31, 121, 170–71; specialties in, 21, 24. *See also* Competition
Physics Letters, 138, 139
Plagiarism, unconscious, 109
Preprints, 89, 104, 131, 133
Prestige hierarchy: effects of, 42–47; measurement of, 41, 175–77; research diversification as factor, 42–43; size as factor, 42; university, 38–47
Price, Derek J. deSolla, 32, 74, 151, 178
Princeton University, 73
Priority disputes, 73, 88–93
Publication, hasty, 72, 105, 106; problem of, 105

Recognition, 3–5, 6, 33, 49–50, 73, 163–66, 173; competition for, 166; correlates of, 53–56; of experimentalists, 64, 159; index, 179–81; measurement of, 49; and scientific productivity, 60–61, 159; of theorists, 64, 159
Reif, F., 3
Replication, 78; of experiments, 112
Research: Department of Scientific and Industrial, 13; Office of Naval, 8, 16; time required for, 61, 162
Research funds: distribution of, 8–9; expenditures, 12–13; source of, 12
Research productivity. *See* Scientific productivity
Reticence, 122, 167
Reward system, 157, 163; in Great Britain, 53, 67; and the nature of science, 59–68; in the United States, 33, 67
Robbins Report, 18, 36, 38, 39, 40
Royal Society, 180, 181; fellow of, 40, 180
Rutherford, Ernest, 4, 22, 70
Rutherford High Energy Laboratory, 17, 45, 80

Science, norms of, 5–7, 73, 173; deviance from, 105; enforcement of, 166

Science and Technology Act of 1965, 13
Science Research Council, 15, 19, 37, 45, 46, 47, 162, 163, 164; composition of, 16; responsibility of, 16; and funds allocated to high energy physics, 17
Scientific American, 85
Scientific conference. *See* Communication
Scientific journals: preference for, 138–39; ranking of, 138; rejection of papers by, 105–7; use of, 132–38. *See also* specific titles
Scientific productivity, 32–34, 47, 160–63; and anticipation, 97; correlates of, 50–53; of experimentalists, 62; index, 177–79; and institutional affiliation, 62–64, 161; measurement of, 48; and professional age, 159, 161; and recognition, 56–59; of theorists, 62; and type of scientist, 62–64, 162
Scientist: experimental, 7; theoretical, 7
Scientists, in study, 10
Secrecy, 72, 105, 116, 117, 138, 167, 174; and anticipation, 116, 117, 159; of experimentalists, 118–21; and stage of research, 122–29; of theorists, 118–21
Selye, Hans, 70
Serpukhov, 26
Shils, Edward, 39
Snow, C. P., 9
Social origins, of high energy physicists, 35–36
Social processes, in science, 69
Social Science Research Council, 15
Social status, of experimental physicists, 31
Soviet bloc, 25
Soviet Union, 25, 80, 174
Spark chambers, 27, 28
State, Secretary of, 13, 14
Storer, Norman, 47
Swatez, Gerald, 23, 78, 79

Technology, Ministry of, 13
Tenure, 165
Theft, 72, 105, 108, 109, 110, 114, 115; of ideas, 107
Thesis topic, selection of, 37–38
Trow, Martin, 7, 35, 39, 40, 161
Truscott, Bruce (pseud.), 39

Undergraduate degrees: first class, 8, 36, 55, 60; second class, 8, 36, 55
Underproductivity, 48
Universities: growth of, 18; number of, 18; Oxbridge, 36, 39, 40, 52, 53, 62, 64, 161; Redbrick, 36, 39, 53, 64; and research opportunities, 62–64, 162; Scottish, 17, 36, 39, 64; in study, 9–10
University Grants Committee, 14, 15, 17, 18, 20, 39, 164; composition of, 14; responsibility of, 14
University of: Aberystwyth, 40; Birmingham, 40; Cambridge, 18, 20, 22, 39, 40; Liverpool, 40; London, 18, 36, 39, 52, 53, 62, 64, 161; Manchester, 40, 161; Nottingham, 40; Oxford, 18, 20, 39, 40; Reading, 161; Stirling, 18; Wales, 36

Visibility, of scientists, 67

Walton, Ernest, 22
Watson, James, 6
Wilson, Mitchell, 71, 123, 127

Zaltman, Gerald, 25, 111, 132
Zloczower, Awraham, 39